Communications and Control Engineering

Springer-Verlag London Ltd.

Published titles include:

Stabilization of Nonlinear Uncertain Systems
Miroslav Krstić and Hua Deng

Passivity-based Control of Euler-Lagrange Systems
Romeo Ortega, Antonio Loría, Per Johan Nicklasson and Hebertt Sira-Ramírez

Stability and Stabilization of Infinite Dimensional Systems with Applications
Zheng-Hua Luo, Bao-Zhu Guo and Omer Morgul

Nonsmooth Mechanics (2nd edition)
Bernard Brogliato

Nonlinear Control Systems II
Alberto Isidori

L_2-Gain and Passivity Techniques in nonlinear Control
Arjan van der Schaft

Control of Linear Systems with Regulation and Input Constraints
Ali Saberi, Anton A. Stoorvogel and Peddapullaiah Sannuti

Robust and H∞ Control
Ben M. Chen

Computer Controlled Systems
Efim N. Rosenwasser and Bernhard P. Lampe

Dissipative Systems Analysis and Control
Rogelio Lozano, Bernard Brogliato, Olav Egeland and Bernhard Maschke

Control of Complex and Uncertain Systems
Stanislav V. Emelyanov and Sergey K. Korovin

Robust Control Design Using H^∞ Methods
Ian R.Petersen, Valery A. Ugrinovski and Andrey V.Savkin

Model Reduction for Control System Design
Goro Obinata and Brian D. O. Anderson

Control Theory for Linear Systems
Harry L. Trentelman, Anton Stoorvogel and Malo Hautus

Functional Adaptive Control
Simon G. Fabri and Visakan Kadirkamanathan

Positive 1D and 2D Systems
Tadeusz Kaczorek

Identification and Control Using Volterra Models
F.J. Doyle III, R.K. Pearson and B.A. Ogunnaike

Isabelle Fantoni and Rogelio Lozano

Non-linear Control for Underactuated Mechanical Systems

With 83 Figures

 Springer

Isabelle Fantoni, Doctor
Rogelio Lozano , Professor

Heudiasyc, UMR CNRS 6599, Universitè de Technologie de Compiègne, BP 20529, 60205 Compiègne, France

Series Editors

E.D. Sontag • M. Thoma

British Library Cataloguing in Publication Data
Fantoni, Isabelle
  Non-linear control for underactuated mechanical systems. -
  (Communications and control engineering)
  1.Nonlinear control theory 2.Automatic control
  3.Mechatronics
  I.Title II.Lozano, R. (Rogelio), 1954-
  629.8'36
  ISBN 978-1-4471-1086-6

Library of Congress Cataloging-in-Publication Data
Fantoni, Isabelle, 1973-
  Non-linear control for underactuated mechanical systems / Isabelle Fantoni and Rogelio
  Lozano .
    p. cm. -- (Communications and control engineering)
  Includes bibliographical references and index.
  ISBN 978-1-4471-1086-6   ISBN 978-1-4471-0177-2 (eBook)
  DOI 10.1007/978-1-4471-0177-2
  1. Automatic control. 2. Nonlinear theories. I. Lozano, R. (Rogelio), 1954- II. Title.
  III. Series.
  TJ213 .L64 2001
  629.8--dc21   2001032030

Communications and Control Engineering Series ISSN 0178-5354

ISBN 978-1-4471-1086-6

Htt://www.springer.co.uk
© Springer-Verlag London 2002
Originally published by Springer-Verlag London Limited in 2002
Softcover reprint of the hardcover 1st edition 2002

Typesetting: Camera ready by authors
Printed and bound at the Cromwell Press, Trowbridge, Wiltshire
69/3830-543210 Printed on acid-free paper SPIN 10780288

Preface

The purpose of this book is to provide a detailed presentation of the control of underactuated non-linear mechanical systems. Control of underactuated systems is a very popular research field, since there exist many applications of underactuated systems in robotics, marine and aerospace vehicles.

Modelling and control of a series of well-known examples of underactuated mechanical systems are presented in this book. The total energy of the system and its passivity properties have been extensively used in the control design. The main goal is the stabilization of controlled dynamical systems by construction of Lyapunov functions. Simulations and real applications illustrate the performance of the algorithms on several experimental platforms.

This book is expected to be used by students and researchers in the areas of non-linear control systems, mechanical systems, robotics and control of helicopters.

The book originates from the Ph.D. thesis prepared by the first author at the University of Technology of Compiègne and supervised by the second author. It also contains four chapters of individual contributions on closely related subjects: Chapter 10 (Carlos Aguilar and Rogelio Lozano), Chapter 13 (Juan Carlos Avila-Vilchis, Bernard Brogliato and Rogelio Lozano), Chapter 14 (Robert Mahony and Rogelio Lozano) and Chapter 15 (Robert Mahony, Tarek Hamel, Alejandro Dzul and Rogelio Lozano).

It would not have been possible to compile the book without the precious help and the contributions of the following persons that are gratefully acknowledged here:

- We are specially indebted to Mark W. Spong, with whom we have closely collaborated under an agreement between CNRS and the University of Illinois. He contributed to the material presented in Chapters 5 and 7.

- We thank Carlos Aguilar, Juan Carlos Avila-Vilchis, Bernard Brogliato, Alejandro Dzul, Tarek Hamel and Robert Mahony for their contributions in Chapters 10 and 13-15 .

- We are grateful to Anuradha M. Annaswamy, Joaquin Collado, Frédéric Mazenc and Kristin Y. Pettersen for fruitful collaboration on the domain of underactuated mechanical systems.

- We also want to thank D. J. Block from the University of Illinois. The experimental results contained in Chapter 3 and in Chapter 5 wouldn't have been possible without his help.

Isabelle Fantoni
Rogelio Lozano

Contents

Chapter 1

Introduction

1.1 Motivation

The motivation for the research presented in this book is to continue the development of the field of non-linear control theory for mechanical systems. The development of robots having an autonomous and complex behavior such as the adaptation to environment changes and uncertainties, planification and execution strategies without human intervention and the learning ability to improve performances is one of the ultimate goals in research of robotics. The achievement of such machines could have a major impact in many fields such as production, stocking and supervision of dangerous waste, construction and the robotic hollow, inspection, teleoperation, maintenance of satellites, autonomous vehicles, etc. It clearly appears that most of the problems to be solved to reach such goals imply control problems. The development of control techniques is a vital objective for the realization and the creation of intelligent robots.

This book presents the application of modern non-linear systems theory to control some important classes of underactuated mechanical systems. In the eighties, the control of robot manipulators was extensively studied. Several control strategies based on passivity, Lyapunov theory, feedback linearization, etc. have been developed for the fully actuated case, i.e. systems with the same number of actuators as degrees of freedom. The techniques developed for fully actuated robots do not apply directly to the case of underactuated non-linear mechanical systems. Underactuated mechanical systems or vehicles are systems with fewer independent control actuators than degrees of freedom to be controlled.

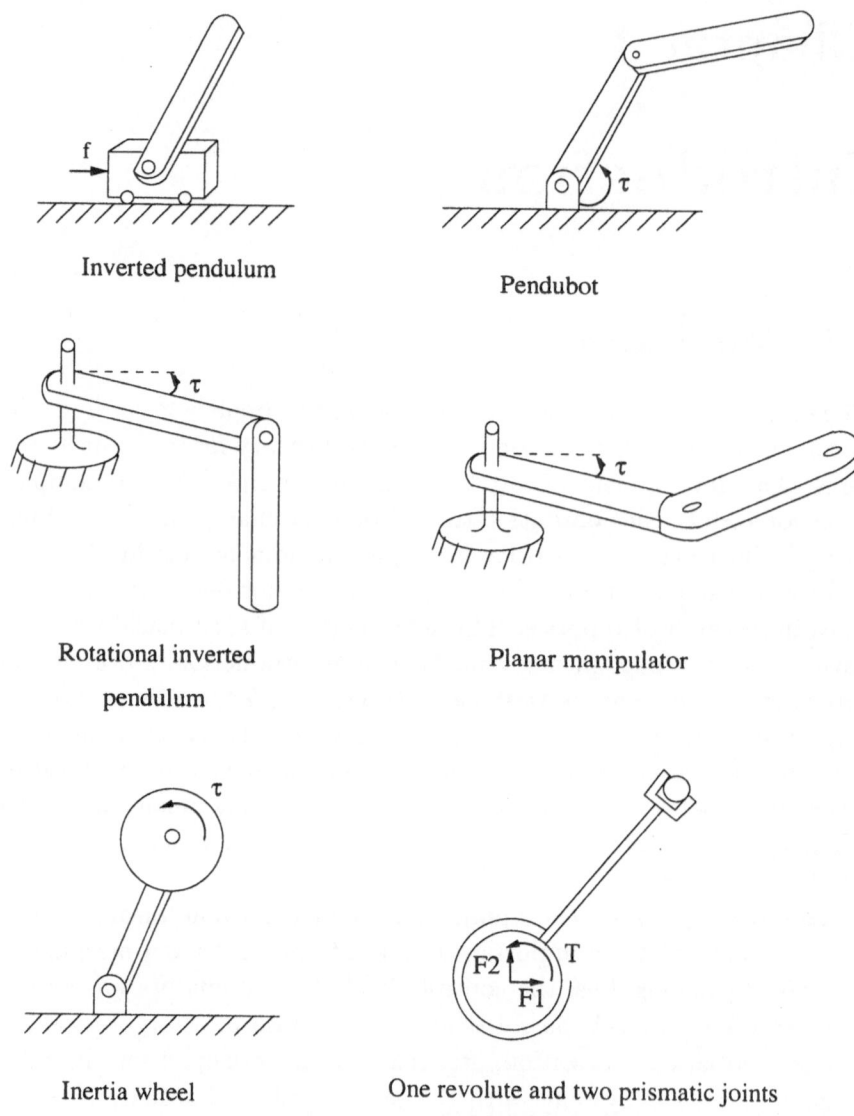

Figure 1.1: Examples of underactuated mechanical systems

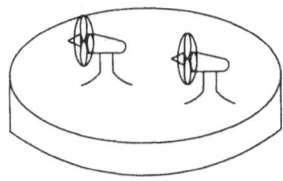

Hovercraft

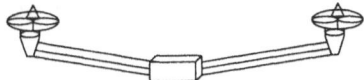

Planar vertical take–off and landing aircraft
(PVTOL)

Figure 1.2: Examples of underactuated mechanical systems

In the last few years, there has been major interest in developing stabilizing algorithms for underactuated mechanical systems. The need for underactuated algorithms arises in many practical situations, some of which are enumerated below. The interest comes from the need to stabilize systems like ships, underwater vehicles, helicopters, aircraft, airships, hovercrafts, satellites, walking robots, etc., which may be underactuated by design. Actuators are expensive and/or heavy and are therefore sometimes avoided in a system design. Other systems may also become underactuated due to actuator failure.

Most models of mechanical systems such as robotic manipulators are built on the assumption that the individual links or members are rigid. This is a correct approximation in some cases while in others it is not. If we take the more realistic non-rigid dynamics into account then all such models are essentially underactuated.

In aircraft or space applications, equipment weight is of paramount importance. Space bound rockets have limited rocket payloads, a sizeable portion of which constitutes automatic control equipment. This has fueled research into underactuated mechanical systems. The idea is simply to reduce the weight of any robotic manipulator by reducing the number of motors, which are often the most heavy and unwieldy parts.

Free floating mobile robots like those used in space and ocean applications often have robotic manipulators mounted on them. Examples include collecting samples in submersible vehicles and performing maintenance operations or debris retrieval in space robots. Close to the objective where the desired task has to be performed, the primary propelling devices are usually switched off so that the platform is free floating. Under such conditions, when the robotics manipulator is moved, the law of conservation dictates that the platform itself will move. So the system becomes underactuated as a new degree of freedom is added and the normal control algorithms do not work.

More specific applications also exist such as in the arena of naval operations, where fuel and rations are supplied to a naval vessel from a supply ship using a swinging crane. Under a high sea state, the relative motion of the two vessels becomes important enough to hinder significantly even the mundane task of loading and unloading. In space applications, if any one of the motors of a multilinked robotic arm malfunction while the arm is extended, the only solution is to jettison the whole assembly. Unless, that is, an underactuated algorithm can be utilized to retrieve it.

Underactuation may be due to an actuator failure. A hardware solution to actuator failures may be achieved by equipping the vehicle with redundant actuators. The software option is, on the other hand, a cost-reducing alternative, since it consists of changing to a control law that controls the vehicle using only the remaining actuators, when an actuator failure is detected. Furthermore, the software solution is weight economical compared to the hardware solution and this can be important in space and underwater applications. Moreover, cost and weight considerations can motivate constructors to create underactuated vehicles.

It is thus clear that there is a need to develop new control techniques applicable to underactuated non-linear mechanical systems.

The research is focused towards obtaining control algorithms for general underactuated non-linear mechanical systems. Since this general objective is difficult to accomplish, we are also interested in stabilizing particular classes of mechanical systems. To simplify the task, researchers have been studying simple mechanical systems. Some of these systems represent academic benchmarks and are part of a standard control laboratory like the inverted pendulum, the rotational inverted pendulum, the pendubot, the planar manipulator with springs between links, the pendulum driven by a spinning wheel, the ball and beam, and the PVTOL (planar vertical take-off and landing) aircraft. In spite of the fact that they are simple mechanical plants, they represent a challenge to the non-linear control community.

The study of the PVTOL aircraft is also important because it represents a simplified model of a helicopter in the lateral (or longitudinal) axis. Indeed, if we consider a helicopter for which the yaw and pitch (or roll) angles are fixed, the resulting system is similar to the PVTOL aircraft. Developing control strategies for the PVTOL aircraft will normally be useful in the control design for helicopters. Chapters 13 through 15 are devoted to modelling and control of helicopter models.

An important remark is the fact that the research on underactuated systems is an extension of the research on non-holonomic systems. Indeed, non-holonomic systems have constraints on the velocity and only kinematic equations of the system are considered. Underactuated systems have constraints on the acceleration, and both kinematics and dynamics have to be considered in the control design.

In this book, we will give a detailed presentation of the control of some well-known underactuated non-linear mechanical systems. The reader will find detailed steps to obtain the Euler-Lagrangian models as well as various control laws obtained using different approaches based on Lyapunov theory, passivity, feedback linearization, etc. Real applications will illustrate the performance of the algorithms on several experimental platforms.

We have neglected friction in the various inverted pendulum systems studied in this book. The convergence analysis proposed deals only with the ideal case in which friction is zero. However, experimental results have shown that the proposed controller performs appropriately when the frictional terms are small.

In most of the examples that will be presented, a passivity- or energy-based approach is used in the design of stabilizing controllers. In fact, the passivity-based control technique is standard and has extensively been studied and used in the control community. Indeed, Willems [122] and Hill & Moylan [37, 38, 39, 40] provided some general notions in the theory of dissipative systems, in particular the small-gain and passivity theorems.

Takegaki & Arimoto [113] developed the idea of stabilizing mechanical systems by reshaping the potential energy via feedback and by adding damping. This is one of the starting points of "passivity-based control".

Byrnes et al. [16] in 1991, derived conditions under which a non-linear system can be rendered passive via smooth state feedback and they extended a number of stabilization schemes for global asymptotic stabilization of certain classes of non-linear systems.

Nijmeijer & Van der Schaft [75] described system theoretic properties and stabilization of standard Hamiltonian control systems. Then Maschke, Van der Schaft & Breedvelt [68] introduced generalized Hamiltonian control systems as an important class of passive state space systems and studied the stabilization of such systems. In 1996, Van der Schaft provided in his lecture notes, published by Springer [117], a very useful synthesis between classical input-output and closed-loop stability theory, in particular the small-gain and passivity theorems, and presented developments in passivity-based and non-linear H_∞ control.

It is also important to remark that recently the journal "International Journal of Robust and Nonlinear Control" has published a special issue entitled "Control of underactuated nonlinear systems". In this issue, a paper by Shiriaev et al. [100] deals with the stabilization of the upright position of the inverted pendulum system that we will develop in Chapter 3. This paper gives an extension on some global properties of the controller that we have proposed in [58] (see also Chapter 3). This clearly shows the great interest that the subject of this book has already received from other researchers of the control community.

The last three chapters of this book deal with modelling and control of helicopters. Helicopters are underactuated systems since they have in general six degrees of freedom (position (x, y, z), pitch, roll and yaw) and only four control inputs (pitching, rolling and yaw moments and the main rotor thrust). Chapter 13 deals with a helicopter mounted on a platform such that the aircraft can move only vertically and around the vertical axis. The system has three degrees of freedom and two inputs. Chapters 14 and 15 deal with helicopters moving freely in a three-dimensional space. Chapter 14 presents a Lagrangian model of the helicopter while Chapter 15 deals with a Newtonian approach. Control laws based on passivity and partial feedback linearization are proposed for each case.

We believe this book will be of great value for Ph.D. students and researchers in the areas of non-linear control systems, mechanical systems, robotics and control of helicopters. It will help to acquire the appropriate models of the proposed systems and to handle other underactuated systems.

1.2   Outline of the book

Chapter 2 presents general notions and background theory, which will be used throughout the book.

1.2.1   Energy-based control approaches for several under-actuated mechanical systems

Chapters 3 to 10 propose a set of underactuated mechanical systems for which we apply an energy-based control approach. For all the systems, we take advantage of their passivity properties in order to establish a control law based on Lyapunov theory. The control objective is to stabilize systems around a desired position. The illustrative examples are the following:

- Chapter 3 deals with the inverted pendulum system and the development of an energy-based control law. The applicability of the method is illustrated by means of simulations and experimental results performed at the University of Illinois at Urbana-Champaign (USA). This work is also published in [59].

- Chapter 4 is a natural extension of Chapter 3. Indeed, the proposed convey-crane system is intensely based on the inverted pendulum's equations. Again, an energy-based control approach is proposed in order to stabilize the convey-crane at its lower equilibrium position. This work has been presented in [18].

- The pendubot system is introduced in Chapter 5. Experiments are also given in order to see the performance of the proposed control law. The presentation of this work can be found in [24].

- In Chapter 6, the rotational inverted pendulum, often called the Furuta pendulum system, is presented and its energy-based control law is developed.

- Chapter 7 deals with the pendulum driven by a spinning wheel, i.e. the reaction wheel. Two different approaches are considered.

- Chapter 8 presents planar underactuated manipulators with springs between links. A simple control law is presented for such systems.

- Chapter 9 introduces a planar robot with two prismatic and one revolute (PPR) joints. This example has four degrees of freedom with only three control inputs. An energy-based control law is again presented.

- Finally, in Chapter 10, we propose a control law for the ball and beam system acting on the ball instead of the beam.

The main contribution of the above systems is to exploit their passivity properties to develop appropriate control laws. Moreover, rigorous and complete stability analysis for the closed-loop systems are presented. Note that the main idea is similar for most of the presented systems. Therefore, it can be extended to a larger class of underactuated mechanical systems provided that the control law is adapted to each particular system.

1.2.2   The hovercraft model, the PVTOL aircraft and the helicopter

From Chapters 11 to 15, we study systems that have a direct real application, such as hovercrafts, aircraft and helicopters. Control laws have been developed using simplified models of such systems.

- Chapter 11 deals with a simplified model of a ship that can also be regarded as a hovercraft model. We propose different control strategies based on Lyapunov theory.

- In Chapter 12, we present the model of a planar vertical take-off and landing (PVTOL) aircraft that is a simplified aircraft. We propose a control strategy by construction of a Lyapunov function using the forwarding technique.

The last three chapters present several methodologies for modelling a helicopter and different control procedures are proposed based on Lyapunov theory.

- Chapter 13 introduces a Lagrangian model of a VARIO scale model helicopter mounted on a platform and a passivity-based control strategy. The proposed control strategy is based on the use of non-linear controllers, which ensure asymptotic tracking of suitable desired trajectories.

- In Chapter 14, a Lagrangian model of the dynamics of a simplified helicopter permits a Lyapunov design of a unified path tracking control algorithm.

- Finally, Chapter 15 presents a Newtonian helicopter model and a robust control design based on robust backstepping techniques is proposed. The control law design is based on an approximation of the system obtained by ignoring the small body forces associated with the torque control.

Chapter 2

Theoretical preliminaries

The purpose of this chapter is to present some definitions and background theory that will be used throughout this book. We will first introduce some important theorems based on Lyapunov theory. We will give some notions and basic concepts of passivity. Then, we will recall a necessary condition for the existence of a continuously stabilizing control law for non-linear systems, often referred to as Brockett's necessary condition. The definitions of non-holonomic systems, under-actuated systems and a homoclinic orbit are also given.

2.1 Lyapunov stability

Consider the autonomous system

$$\dot{x} = f(x) \qquad (2.1)$$

where $f : D \to I\!\!R^n$ is a locally Lipschitz map from a domain $D \subset I\!\!R^n$ into $I\!\!R^n$. We suppose that the origin $x = 0$ is an equilibrium point of (2.1), which satisfies

$$f(0) = 0$$

Lyapunov theory is the fundamental tool for stability analysis of dynamic systems. The following definitions and theorems are used to characterize and study the stability of the origin (see Khalil [46]).

Definition 2.1 *(Khalil, 1996, Definition 3.1)* The equilibrium point $x = 0$ of system (2.1) is

- stable, if for each $\epsilon > 0$, there is $\delta = \delta(\epsilon) > 0$ such that

$$\|x(0)\| < \delta \Rightarrow \|x(t)\| < \epsilon, \quad \forall t \geq 0$$

- unstable, if not stable

- asymptotically stable, if it is stable and δ can be chosen such that

$$\|x(0)\| < \delta \Rightarrow \lim_{t \to \infty} x(t) = 0$$

- exponentially stable, if there exist two strictly positive numbers α and λ independent of time and initial conditions such that

$$\|x(t)\| \leq \alpha \|x(0)\| \exp(-\lambda t) \qquad \forall t > 0 \qquad (2.2)$$

  in some ball around the origin.   ∎

The above definitions correspond to *local* properties of the system around the equilibrium point. The above stability concepts become *global* when their corresponding conditions are satisfied for *any initial state*.

2.1.1   Lyapunov direct method

Let us consider the following definitions.

Definition 2.2 ((Semi-)definiteness) A scalar continuous function $V(x)$ is said to be locally *positive (semi-) definite* if $V(0) = 0$ and $V(x) > 0$ $(V(x) \geq 0)$ for $x \neq 0$. Similarly, $V(x)$ is said to be *negative (semi-)definite* if $-V(x)$ is positive (semi-)definite.   ∎

Definition 2.3 (Lyapunov function) $V(x)$ is called a *Lyapunov function* for the system (2.1) if, in a ball B containing the origin, $V(x)$ is positive definite and has continuous partial derivatives, and if its time derivative along the solutions of (2.1) is negative semi-definite, i.e. $\dot{V}(x) = (\partial V/\partial x)f(x) \leq 0$.   ∎

The following theorems can be used for local and global analysis of stability, respectively.

Theorem 2.1 *(Local stability)* *The equilibrium point 0 of system (2.1) is (asymptotically) stable in a ball B if there exists a scalar function $V(x)$ with continuous derivatives such that $V(x)$ is positive definite and $\dot{V}(x)$ is negative semi-definite (negative definite) in the ball B.*   ∎

Theorem 2.2 *(Global stability)* *The equilibrium point of system (2.1) is globally asymptotically stable if there exists a scalar function $V(x)$ with continuous first order derivatives such that $V(x)$ is positive definite, $\dot{V}(x)$ is negative definite and $V(x)$ is radially unbounded, i.e. $V(x) \to \infty$ as $\|x\| \to \infty$.* ∎

Krasovskii-LaSalle's invariant set theorem

Krasovskii-LaSalle's results extend the stability analysis of the previous theorems when $\dot{V}$ is only negative semi-definite. They are stated as follows.

Definition 2.4 (Invariant set) A set S is an *invariant set* for a dynamic system if every trajectory starting in S remains in S. ∎

Invariant sets include equilibrium points, limit cycles, as well as any trajectory of an autonomous system.

Theorem 2.3 *(LaSalle's invariance principle)* *(Khalil, 1996, Theorem 3.4) Let Ω be a compact (closed and bounded) set with the property that every solution of the system (2.1) that starts in Ω remains in Ω for all future time. Let $V : \Omega \to I\!\!R$ be a continuously differentiable function such that $\dot{V}(x) \leq 0$ in Ω. Let E be the set of all points in Ω where $\dot{V}(x) = 0$. Let M be the largest invariant set in E. Then, every solution starting in Ω approaches M as $t \to \infty$.* ∎

When $V(x)$ is positive definite, the following two corollaries extend Theorems 2.1 and 2.2.

Corollary 2.1 *(Barbashin-LaSalle)* *(Khalil, 1996, Corollary 3.1) Let $x = 0$ be an equilibrium point for (2.1). Let $V : D \to I\!\!R$ be a continuously differentiable positive definite function on a neighborhood D of $x = 0$, such that $\dot{V}(x) \leq 0$ in D. Let $S = \{x \in D | \dot{V}(x) = 0\}$, and suppose that no solution can stay forever in S, other than the trivial solution. Then, the origin is asymptotically stable.* ∎

Corollary 2.2 *(Krasovskii-LaSalle)* *(Khalil, 1996, Corollary 3.2) Let $x = 0$ be an equilibrium point for (2.1). Let $V : I\!\!R^n \to I\!\!R$ be a continuously differentiable, radially unbounded, positive definite function such that $\dot{V}(x) \leq 0$ for all $x \in I\!\!R^n$. Let $S = \{x \in I\!\!R^n | \dot{V}(x) = 0\}$, and suppose that no solution can stay forever in S, other than the trivial solution. Then, the origin is globally asymptotically stable.* ∎

When $\dot{V}(x)$ is negative definite, $S = \{0\}$. Then, Corollaries 2.1 and 2.2 coincide with Theorems 2.1 and 2.2, respectively.

2.2   Passivity and dissipativity

Consider the non-linear system

$$\begin{aligned}
\dot{\mathbf{x}} &= f(\mathbf{x}) + g(\mathbf{x})u \\
\mathbf{y} &= h(\mathbf{x}) + j(\mathbf{x})u
\end{aligned} \qquad (2.3)$$

$\mathbf{x} \in \mathbb{R}^n$, $u, y \in \mathbb{R}^m$, f, g, h, j are smooth. $f(0) = h(0) = 0$ (see Lozano [56]). Let us call $w = w(u, y)$ the supply rate, such that $\forall u, \forall x(0)$

$$\int_0^t |w(s)|\,ds < \infty \qquad\qquad t \in \mathbb{R}^+$$

i.e. locally integrable.

Definition 2.5 (Dissipative system) The system (2.3) is said to be dissipative if there exists a storage function $V(x) \geq 0$ such that $\forall u, \forall x(0)$

$$V(x(t)) - V(x(0)) \leq \int_0^t w(s)ds$$

The latter is called a dissipation inequality.   ∎

It means that the storage energy function $V(x(t))$ at a future time t is not bigger than the sum of the available storage function $V(x(0))$ at an initial time 0 plus the total energy $\int_0^t w(s)ds$ supplied to the system from the external sources in the interval $[0, t]$. There is no internal creation of energy.

Passive systems represent an important subset of dissipative systems.

Definition 2.6 A system with input u and output y where $u(t), y(t) \in \mathbb{R}^n$ is passive if there is a constant β such that

$$\int_0^T y^T(t)u(t)dt \geq \beta \qquad (2.4)$$

for all functions u and all $T \geq 0$. If, in addition, there are constants $\delta \geq 0$ and $\epsilon \geq 0$ such that

$$\int_0^T y^T(t)u(t)dt \geq \beta + \delta \int_0^T u^T(t)u(t)dt + \epsilon \int_0^T y^T(t)y(t)dt \quad (2.5)$$

for all functions u, and all $T \geq 0$, then the system is input strictly passive if $\delta > 0$, output strictly passive if $\epsilon > 0$, and very strictly passive if $\delta > 0$ and $\epsilon > 0$ ■

Obviously $\beta \leq 0$ as the inequality is to be valid for all functions u and in particular the control $u(t) = 0$ for all $t \geq 0$, which gives $0 = \int_0^T y^T(t)u(t)dt \geq \beta$.

Theorem 2.4 *Assume that there is a continuous function $V(t) \geq 0$ such that*

$$V(T) - V(0) \leq \int_0^T y(t)^T u(t)dt \quad (2.6)$$

for all functions u, for all $T \geq 0$ and all $V(0)$. Then, the system with input $u(t)$ and output $y(t)$ is passive. Assume, in addition, that there are constants $\delta \geq 0$ and $\epsilon \geq 0$ such that

$$V(T) - V(0) \leq \int_0^T y^T(t)u(t)dt - \delta \int_0^T u^T(t)u(t)dt - \epsilon \int_0^T y^T(t)y(t)dt$$
$$(2.7)$$

for all functions u, for all $T \geq 0$ and all $V(0)$. Then, the system is input strictly passive if $\delta > 0$, output strictly passive if $\epsilon > 0$, and very strictly passive if $\delta > 0$ and $\epsilon > 0$ such that the inequality holds. ■

2.3 Stabilization

The following theorem gives a necessary condition for the existence of a continuously differentiable control law for non-linear systems. It was presented by Brockett [13] (1983) for C^1 pure-state feedback laws.

Theorem 2.5 *Let $\dot{x} = f(x, u)$ be given with $f(x_0, 0) = 0$ and $f(., .)$ continuously differentiable in a neighborhood of $(x_0, 0)$. A necessary condition for the existence of a continuously differentiable control law which makes $(x_0, 0)$ asymptotically stable is that:*

(i) *The linearized system should have no uncontrollable modes associated with eigenvalues whose real part is positive.*

(ii) *There exists a neighborhood N of $(x_0, 0)$ such that for each $\zeta \in N$ there exists a control $u_\zeta(.)$ defined on $[0, \infty[$ such that this control steers the solution of $\dot{x} = f(x, u_\zeta)$ from $x = \zeta$ at $t = 0$ to $x = x_0$ at $t = \infty$.*

(iii) *The mapping $\gamma : A \times I\!\!R^m \to I\!\!R^n$ defined by $\gamma : (x, u) \mapsto f(x, u)$ should be onto an open set containing 0.*   ■

The first condition refers to the rank condition of a linear control system. Note that in the linear case, the rank condition is necessary and sufficient for a linear system $\dot{x} = Ax + Bu$ to be controllable and to provide the existence of a continuously differentiable control law for the linear system.

The second condition refers to the controllability property in the nonlinear case. On the other hand, this condition is not sufficient since we want a control law with some smoothness. In general, we need something more than just a controllability condition. Therefore, we need to introduce condition (iii), which corresponds to the necessary condition of this theorem.

The third condition means that the mapping should be locally surjective or that the image of the mapping $(x, u) \mapsto f(x, u)$, for x and u arbitrarily close to 0, should contain a neighborhood of the origin.

To make this more clear, let us consider the following example:

$$\begin{aligned} \dot{x} &= u = \varepsilon_1 \\ \dot{y} &= v = \varepsilon_2 \\ \dot{z} &= yu - xv = \varepsilon_3 \end{aligned}$$

The question to ask is if there exists a continuous control law $(u, v) = (u(x, y, z), v(x, y, z))$ that makes the origin asymptotically stable for the above system. The third condition of Brockett's theorem means that the system equations should contain a solution (x, y, z, u, v) for every $\varepsilon_i (i = 1, 2, 3)$ in a neighborhood of the origin. This is not the case here, since the system does not have a solution for $\varepsilon_3 \neq 0$ and $\varepsilon_1 = 0$, $\varepsilon_1 = 0$.

Contrary to the above, the following example satisfies the third condition

$$\begin{aligned} \dot{x} &= u = \varepsilon_1 \\ \dot{y} &= v = \varepsilon_2 \\ \dot{z} &= xy = \varepsilon_3 \end{aligned}$$

Therefore, for this particular system there exists a continuous control law, which makes the origin asymptotically stable.

2.4 Non-holonomic systems

Definition 2.7 (Holonomic systems) (Goldstein [33]) Consider a system of generalized coordinates q

$$\ddot{q} = f(q, \dot{q}, u) \tag{2.8}$$

where $f(.)$ is the vector field representing the dynamics and u is a vector of external generalized inputs. Suppose that some constraints limit the motion of the system. If the conditions of constraint can be expressed as equations connecting the coordinates (and possibly the time) having the form

$$h(q, t) = 0 \tag{2.9}$$

then the constraints are said to be *holonomic*. This type of constraint is a so-called holonomic constraint, since it can be integrated. ■

A simple and suitable example illustrates well the concept of holonomic systems and was proposed by Lefeber [50]. Let us consider the system

$$
\begin{aligned}
\dot{x}_1 &= u x_2 \\
\dot{x}_2 &= -u x_1
\end{aligned}
\tag{2.10}
$$

where (x_1, x_2) is the state and u is the input. This system contains a constraint on the velocities as follows

$$x_1 \dot{x}_1 + x_2 \dot{x}_2 = 0 \tag{2.11}$$

Since this constraint can be integrated to obtain

$$\frac{1}{2} x_1^2 + \frac{1}{2} x_2^2 = c \tag{2.12}$$

where c is a constant, the constraint (2.11) is called a holonomic constraint.

Definition 2.8 (Non-holonomic systems) (Goldstein [33]). On the other hand, when it is not possible to reduce them further by means of equations of constraint of the form (2.9), they are then called *non-holonomic*. With non-holonomic systems, the generalized coordinates are not independent of each other. ■

Let us consider the kinematic equations of a vehicle

$$\begin{aligned} \dot{x} &= v\cos\theta \\ \dot{y} &= v\sin\theta \end{aligned} \tag{2.13}$$

The constraint on the velocities of the model is given by

$$\dot{x}\sin\theta - \dot{y}\cos\theta = 0 \tag{2.14}$$

However, contrary to (2.11), the constraint (2.14) cannot be integrated, i.e. the constraint (2.14) cannot be written as a time derivative of some function of the state. It is called a non-holonomic constraint.

2.5   Underactuated systems

In this book, we will simply define an underactuated system as one having less control inputs than degrees of freedom. The precise definition is given below. In some underactuated systems, the lack of actuation on certain directions can be interpreted as constraints on the acceleration. This is in fact the case for the underactuated hovercraft, treated in Chapter 11, for which the lateral acceleration is nil.

Definition 2.9 (Underactuated systems) Consider systems that can be written as

$$\ddot{q} = f(q, \dot{q}) + G(q)u \tag{2.15}$$

where q is the state vector of independent generalized coordinates, $f(.)$ is the vector field representing the dynamics of the systems, $\dot{q}$ is the generalized velocity vector, G is the input matrix, and u is a vector of generalized force inputs. The dimension of q is defined as the degrees of freedom of (2.15). System (2.15) is said to be underactuated if the external generalized forces are not able to command instantaneous accelerations in all directions in the configuration space, i.e. $rank(G) < dim(q)$.
∎

This definition is connected to the one used by Oriolo and Nakamura [80], which says that underactuated systems are systems with fewer independent control actuators than degrees of freedom <u>to be controlled</u>.

Several examples of underactuated systems will be considered in this book. In order to illustrate the above definition, let us consider the model of the pendubot system (see Figure 5.1), which will be developed in Chapter 5. The dynamic equations of the system in standard form are given by

$$D(q)\ddot{q} + C(q, \dot{q})\dot{q} + g(q) = \tau \tag{2.16}$$

where

$$q = \begin{bmatrix} q_1 \\ q_2 \end{bmatrix} \qquad D(q) = \begin{bmatrix} \theta_1 + \theta_2 + 2\theta_3 \cos q_2 & \theta_2 + \theta_3 \cos q_2 \\ \theta_2 + \theta_3 \cos q_2 & \theta_2 \end{bmatrix} \tag{2.17}$$

$$C(q, \dot{q}) = \begin{bmatrix} -\theta_3 \sin(q_2)\,\dot{q}_2 & -\theta_3 \sin(q_2)\,\dot{q}_2 - \theta_3 \sin(q_2)\,\dot{q}_1 \\ \theta_3 \sin(q_2)\,\dot{q}_1 & 0 \end{bmatrix} \tag{2.18}$$

$$g(q) = \begin{bmatrix} \theta_4 g \cos q_1 + \theta_5 g \cos(q_1+q_2) \\ \theta_5 g \cos(q_1+q_2) \end{bmatrix} \quad \text{and} \quad \tau = \begin{bmatrix} \tau_1 \\ 0 \end{bmatrix} \tag{2.19}$$

In this system, there is only one actuator acting on the first link (i.e. the angle q_1), while the second link (i.e. the angle q_2) is free. Indeed, in the vector τ, there is only one term τ_1 on the first line. Therefore, this system is underactuated since it has two degrees of freedom with only one actuator.

Remark 2.1 *Let us note that for the case of the mobile robot in kinematic equations (2.13) the generalized coordinates are (x, y, θ) and are of three-dimensional while the system has two control inputs (the forward acceleration and the angular momentum). Thus, the mobile robot is an underactuated system in view of Definition 2.9. Actually as shown in equation (2.14), the generalized coordinate components are not independent. This comes from the fact that the system cannot move laterally. Furthermore, an actuator to move the system laterally is beside the point. Note, however, that in this case there are only two degrees of freedom <u>to be controlled</u>: the forward (or backward) displacement and the angular position. In some sense, the system is fully actuated. This shows the limitations of Definition 2.9.* ∎

2.6 Homoclinic orbit

The notions and examples developed in this section are related to the book of Jackson [44]. Let us consider autonomous systems of the form

$$\dot{x} = F(x; c) \qquad x \in \mathbb{R}^n \qquad \text{(autonomous)} \tag{2.20}$$

The solutions of (2.20) involving different initial conditions generate the *family of oriented phase curves* in the phase space, which is called the *phase portrait* of the system. The phase portrait consists of oriented curves through all points of the phase space, where the functions $F(x; c)$ in (2.20) are defined. These curves are called *trajectories* or *orbits*.

One important example is the *limit cycle*, which is a closed (periodic) orbit, whose neighboring orbits tend asymptotically towards (or away) from it.

Definition 2.10 (Homoclinic orbit) A *homoclinic orbit* is a single orbit where a stable manifold and an unstable manifold intersect. This orbit leaves the saddle point in one direction, and returns in another direction. It converges to the same saddle point.   ∎

The example of a ball rolling without friction on a curved surface with maxima of different heights is a simple example of a homoclinic orbit. In this example, there is a single orbit that converges to the same saddle point both when $t \to +\infty$ and when $t \to -\infty$. This situation is illustrated in Figure 2.1, where the height of the surface is shown in (a).

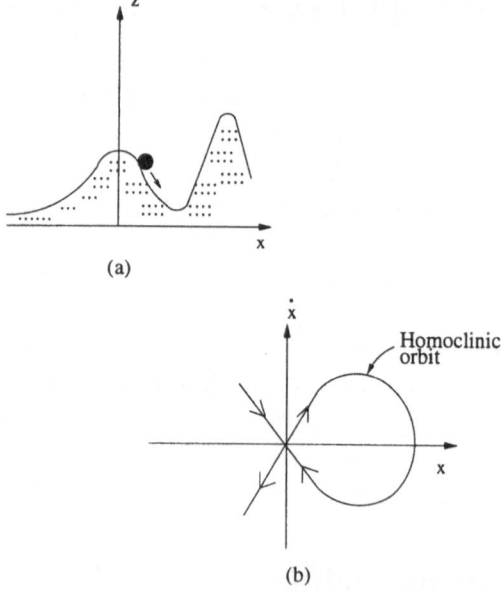

Figure 2.1: Example of a homoclinic orbit

Chapter 3

The cart-pole system

3.1 Introduction

The inverted pendulum is one of the most popular laboratory experiments used for illustrating non-linear control techniques. This system is motivated by applications such as the control of rockets and the anti-seismic control of buildings.

The swinging pendulum on a cart consists of a pole whose pivot point is mounted on a cart, which is a movable platform. The pendulum is free to swing about its pivot point and it has no direct control actuation. The cart can move horizontally perpendicular to the axis of rotation of the pendulum and is actuated by a force applied to it in the same direction. The control objective is to bring the pole to the upper unstable equilibrium position by moving the cart on the horizontal plane. Since the angular acceleration of the pole cannot be controlled directly, the inverted pendulum is an underactuated mechanical system. Therefore, the techniques developed for fully actuated mechanical robot manipulators cannot be used to control the inverted pendulum.

The cart-pole system is also known because several of its properties prohibit the use of standard non-linear control techniques and make it an interesting research problem. Indeed, the relative degree [42] of the system is not constant (when the output is chosen to be the swinging energy of the pendulum), which means that the system is not input-output linearizable. Moreover, Jakubczyk and Respondek [45] have shown that the inverted pendulum is not feedback linearizable. An additional difficulty comes from the fact that when the pendulum swings past the horizontal, the controllability distribution does not have a constant rank and so the system loses controllability as the pendulum swings past its horizontal configuration.

Different interesting controllers exist in the literature. Wei et al. [121] presented a control strategy decomposed in a sequence of steps to bring the pendulum from its lower stable equilibrium position to its unstable equilibrium position, when the cart has a restricted horizontal travel. Chung and Hauser [17] proposed a non-linear state feedback control law to regulate the cart position as well as the swinging energy of the pendulum. The resulting closed-loop system possesses a locally stable periodic orbit, though the region of attraction has not been determined. Lin et al. [55] proposed a linear controller that stabilizes the linearized model of the inverted pendulum, having restricted travel. The region of attraction when the controller is applied to the non-linear model of the inverted pendulum is still to be determined. Fradkov [27] proposed a swinging control strategy of non-linear oscillations. His approach can, in particular, be applied to stabilize an inverted pendulum. Another interesting approach to swing up a pendulum by energy control is given by Åström and Furuta [4]. Note that in [27] and [4], the model does not include the cart displacement. Mazenc and Praly [72] presented a control law based on the technique consisting of adding integrators. Their technique can be used to stabilize the inverted pendulum in its upper equilibrium position when the pendulum is initially above the horizontal plane. Contrary to other strategies, their approach is such that the cart displacement converges to zero. Praly [86], Spong and Praly [109] proposed a strategy to control the inverted pendulum by swinging it up to its unstable equilibrium position. The stability analysis is carried out by using a Lyapunov technique. Note that as a starting point, they used a simplified system that results from the application of partial feedback linearization. In 1999, Olfati-Saber [77] considered stabilization of a special class of cascade non-linear systems consisting of a non-linear subsystem in cascade with a double integrator system. He developed fixed point backstepping procedures for global and semi-global stabilization of this special class of cascade non-linear systems. He demonstrated a reduction strategy by applying his theoretical results to stabilization of the cart-pole system to a point equilibrium over the upper half plane. Semi-global stabilization is achieved using fixed point controllers.

The stabilization algorithm that we propose, is inspired by the work in [86] and [109]. The system stability is likewise demonstrated using Lyapunov analysis but, in contrast to the last authors, the controller is designed directly without partial feedback linearization. The difference relies also on the fact that in the present approach, the control algorithm is obtained by considering the total energy of the inverted pendulum. The inherent non-linearities of the system are not canceled before the control design. This simplifies the closed-loop stability analysis and renders the technique potentially applicable to a wider class of underactuated mechanical systems like the pendubot (see [24] and Chapter 5), the planar manipulators with springs between the links using a single actuator (see [21] and Chapter 8) and the Furuta pendulum (see Chapter 6). The control algorithm as well as the convergence analysis turns out to be very simple as compared to the existing control strategies.

This chapter is organized as follows. In Section 3.2, the model of the inverted pendulum system is given. Section 3.3 presents the passivity properties of the system. In Section 3.4, the controllability of the linearized system is studied. Section 3.5 deals with the stabilization of the system around its homoclinic orbit using an energy approach. Section 3.6 presents the stability analysis of the proposed control law. The performance of the control law is exposed in a simulation example and in real-time experiments, in Sections 3.7 and 3.8. Finally, Section 3.9 gives some conclusions and remarks.

3.2 Model derivation

In this section, the mathematical model of the cart and pendulum system as shown in Figure 3.1 is derived using both Newton's second law and the Euler-Lagrange formulation. We will consider the standard assumptions, i.e. no friction, no dissipative forces, etc.

M	:	Mass of the cart
m	:	Mass of the pendulum
l	:	Distance from the pivot point to the center
	:	of gravity of the pendulum
I	:	Inertia of the pendulum about its center of gravity
g	:	Acceleration due to gravity
x	:	Distance of the cart's center of mass from its initial
	:	position
θ	:	Angle that the pendulum makes with the vertical
f	:	Force applied on the cart

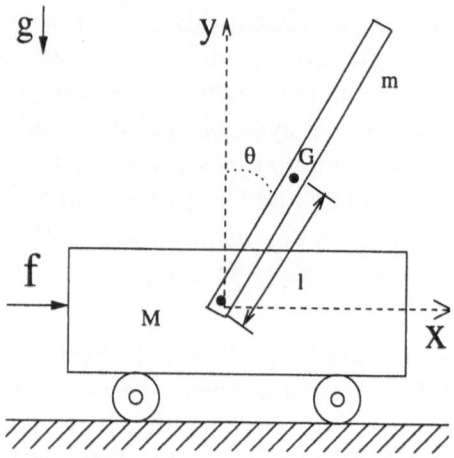

Figure 3.1: The cart pendulum system

3.2.1   System model using Newton's second law

The coordinates of the pendulum's center of mass (x_G, y_G) are

$$x_G = x + l \sin \theta$$
$$y_G = l \cos \theta$$

Applying Newton's second law in the x direction we get

$$
\begin{aligned}
f &= M \frac{d^2 x}{dt^2} + m \frac{d^2 x_G}{dt^2} \\
&= M \frac{d^2 x}{dt^2} + m \frac{d^2}{dt^2} \left(x + l \sin \theta \right) \\
&= M \ddot{x} + m \left(\ddot{x} + l \frac{d}{dt} \cos \theta \dot{\theta} \right) \\
&= M \ddot{x} + m \ddot{x} + ml \left(\cos \theta \ddot{\theta} - \sin \theta \dot{\theta}^2 \right) \\
&= (M + m) \ddot{x} - ml \left(\sin \theta \right) \dot{\theta}^2 + ml \left(\cos \theta \right) \ddot{\theta} \qquad (3.1)
\end{aligned}
$$

Let us now apply Newton's second law to the rotational motion. Recall that the moment of inertia of the pendulum is I. The angular momentum of the pendulum M_A is composed of terms involving rotation about a fixed axis. While the pendulum rotates about its center of mass, the angular momentum is simply $I\dot\theta$. Since the pendulum rotates about the pivot point, a second term due to the distance between the center of mass and the pivot appears. Therefore the angular momentum becomes $M_A = (I + ml^2)\dot\theta$.

The rotational motion of the pendulum involves two forces: the force due to gravity and the force due to the acceleration of the cart. Indeed, since the cart is moving, it applies a force on the pendulum. The moment of the force due to gravity is $mgl\sin\theta$ and the moment of the force due to the acceleration of the cart is $-m\ddot{x}l\cos\theta$.

Newton's second law states that the time derivative of the angular momentum is equal to the moment of the forces applied on the system. Therefore, we obtain

$$(I + ml^2)\ddot\theta = mgl\sin\theta - ml\ddot{x}\cos\theta \qquad (3.2)$$

Finally, Equations (3.1) and (3.2) describe the dynamic behavior of the system.

3.2.2 Euler-Lagrange's equations

We first present the kinetic and potential energies, which are used to compute the Lagrangian function. The kinetic energy of the cart is $K_1 = \frac{M\dot{x}^2}{2}$. The kinetic energy of the pendulum is $K_2 = \frac{m\dot{x}_G^2}{2} + \frac{m\dot{y}_G^2}{2} + \frac{I\dot\theta^2}{2}$ where $x_G = x + l\sin\theta$ and $y_G = l\cos\theta$. The total kinetic energy is then

$$K = K_1 + K_2 = \frac{1}{2}(M + m)\dot{x}^2 + ml\dot{x}\dot\theta\cos\theta + \frac{1}{2}(I + ml^2)\dot\theta^2$$

The total potential energy is $P = mgl(\cos\theta - 1)$. The Lagrangian function is given by

$$L = K - P$$
$$L = \frac{1}{2}(M + m)\dot{x}^2 + ml\dot{x}\dot\theta\cos\theta + \frac{1}{2}(I + ml^2)\dot\theta^2 - mgl(\cos\theta - 1)$$

The corresponding equations of motion are derived using Lagrange's equations

$$\frac{d}{dt}\left(\frac{\partial L}{\partial \dot{q}}(q, \dot{q})\right) - \frac{\partial L}{\partial q}(q, \dot{q}) = \tau \qquad (3.3)$$

where $q = (q_1, ... q_n)^T$ represents the generalized variables, one for each degree of freedom of the system, $\tau = (\tau_1, ..., \tau_n)^T$ denotes forces that are externally applied to the system.

In our case, the generalized variables are x and θ, i.e. $q = (x, \theta)^T$. We therefore have

$$\left(\frac{\partial L}{\partial \dot{x}}\right)^{\bullet} = (M + m)\ddot{x} + ml\dot{\theta} \cos \theta$$

$$\left(\frac{\partial L}{\partial x}\right) = 0$$

$$\left(\frac{\partial L}{\partial \dot{\theta}}\right)^{\bullet} = ml\dot{x} \cos \theta + (I + ml^2)\dot{\theta}$$

$$\left(\frac{\partial L}{\partial \theta}\right) = mgl \sin \theta - ml\dot{x}\dot{\theta} \sin \theta$$

From Lagrange's equations (3.3), we finally obtain the equations of motion (3.1) and (3.2).

In the following, we will assume that the inertia of the pendulum is negligible, so that we will cancel it from Equations (3.1) and (3.2). Note that it could be included in the model and in the control law.

The system can be written in standard form

$$M(q)\ddot{q} + C(q, \dot{q})\dot{q} + G(q) = \tau \qquad (3.4)$$

where

$$q = \begin{bmatrix} x \\ \theta \end{bmatrix} \quad M(q) = \begin{bmatrix} M + m & ml \cos \theta \\ ml \cos \theta & ml^2 \end{bmatrix} \qquad (3.5)$$

$$C(q, \dot{q}) = \begin{bmatrix} 0 & -ml \sin \theta \dot{\theta} \\ 0 & 0 \end{bmatrix} \qquad (3.6)$$

$$G(q) = \begin{bmatrix} 0 \\ -mgl \sin \theta \end{bmatrix} \quad \text{and} \quad \tau = \begin{bmatrix} f \\ 0 \end{bmatrix} \qquad (3.7)$$

Note that $M(q)$ is symmetric and

$$\begin{aligned} \det(M(q)) &= (M + m)ml^2 - m^2l^2\cos^2\theta \\ &= Mml^2 + m^2l^2\sin^2\theta > 0 \end{aligned} \qquad (3.8)$$

Therefore, $M(q)$ is positive definite for all q. From (3.5) and (3.6), it follows that

$$\dot{M} - 2C = \begin{bmatrix} 0 & ml\sin\theta\dot{\theta} \\ -ml\sin\theta\dot{\theta} & 0 \end{bmatrix} \tag{3.9}$$

which is a skew-symmetric matrix. An important property of skew-symmetric matrices, which will be used in establishing the passivity property of the inverted pendulum, is

$$z^T(\dot{M}(q) - 2C(q,\dot{q}))z = 0 \quad \forall z \in \mathbb{R}^2 \tag{3.10}$$

The potential energy of the pendulum can be defined as $P = mgl(\cos\theta - 1)$. Note that P is related to $G(q)$ as follows

$$G(q) = \frac{\partial P}{\partial q} = \begin{bmatrix} 0 \\ -mgl\sin\theta \end{bmatrix} \tag{3.11}$$

3.3 Passivity of the inverted pendulum

The total energy of the cart-pole system is given by

$$\begin{aligned} E &= K(q,\dot{q}) + P(q) \\ &= \tfrac{1}{2}\dot{q}^T M(q)\dot{q} + mgl(\cos\theta - 1) \end{aligned} \tag{3.12}$$

Therefore, from (3.4)-(3.6), (3.7), (3.9)-(3.11) we obtain

$$\begin{aligned} \dot{E} &= \dot{q}^T M(q)\ddot{q} + \tfrac{1}{2}\dot{q}^T \dot{M}(q)\dot{q} + \dot{q}^T G(q) \\ &= \dot{q}^T(-C\dot{q} - G + \tau + \tfrac{1}{2}\dot{M}\dot{q}) + \dot{q}^T G \\ &= \dot{q}^T\tau = \dot{x}f \end{aligned} \tag{3.13}$$

Integrating both sides of the above equation, we obtain

$$\begin{aligned} \int_0^t \dot{x}f\,dt &= E(t) - E(0) \\ &\geq -2mgl - E(0) \end{aligned} \tag{3.14}$$

Therefore, the system having f as input and $\dot{x}$ as output is passive. Note that for $f = 0$ and $\theta \in [0, 2\pi[$, the system (3.4) has a subset of two equilibrium points; $(x, \dot{x}, \theta, \dot{\theta}) = (*, 0, 0, 0)$ is an unstable equilibrium point and $(x, \dot{x}, \theta, \dot{\theta}) = (*, 0, \pi, 0)$ is a stable equilibrium point. The total energy $E(q, \dot{q})$ is equal to 0 for the unstable equilibrium point and to $-2mgl$ for the stable equilibrium point. The control objective is to stabilize the system around its unstable equilibrium point, i.e. to bring the pendulum to its upper position and the cart displacement to zero simultaneously.

3.4   Controllability of the linearized model

When the pendulum is in a neighborhood of its top unstable equilibrium position, it is a well-known fact that a linear controller can satisfactorily stabilize the pendulum. In order to implement a balancing linear controller, the general non-linear differential equations (3.4) have to be linearized about the top equilibrium position and the resulting system has to be controllable. Let us therefore compute the rank of the controllability matrix. The general non-linear equations are given by (see (3.4))

$$
\begin{aligned}
\ddot{x} &= \tfrac{1}{M+m\sin^2\theta}\left[m\sin\theta(l\dot{\theta}^2 - g\cos\theta) + f\right] \\
\ddot{\theta} &= \tfrac{1}{l(M+m\sin^2\theta)}\left[-ml\dot{\theta}^2\sin\theta\cos\theta + (M+m)g\sin\theta - f\cos\theta\right]
\end{aligned}
\tag{3.15}
$$

Linearizing the non-linear equations about the top unstable equilibrium point, we obtain as a resulting linear system

$$
\frac{d}{dt}\begin{bmatrix} x \\ \dot{x} \\ \theta \\ \dot{\theta} \end{bmatrix} = \begin{bmatrix} 0 & 1 & 0 & 0 \\ 0 & 0 & -\frac{mg}{M} & 0 \\ 0 & 0 & 0 & 1 \\ 0 & 0 & \frac{(M+m)g}{lM} & 0 \end{bmatrix}\begin{bmatrix} x \\ \dot{x} \\ \theta \\ \dot{\theta} \end{bmatrix} + \begin{bmatrix} 0 \\ \frac{1}{M} \\ 0 \\ -\frac{1}{lM} \end{bmatrix} f = AX + Bf
$$

with obvious notation. We then have

$$
B = \begin{bmatrix} 0 \\ \frac{1}{M} \\ 0 \\ -\frac{1}{lM} \end{bmatrix}, \quad AB = \begin{bmatrix} \frac{1}{M} \\ 0 \\ -\frac{1}{lM} \\ 0 \end{bmatrix}, \quad A^2B = \begin{bmatrix} 0 \\ \frac{mg}{M^2l} \\ 0 \\ -\frac{(M+m)g}{l^2M^2} \end{bmatrix},
$$

$$A^3 B = \begin{bmatrix} \frac{mg}{M^2 l} \\ 0 \\ -\frac{(M+m)g}{l^2 M^2} \\ 0 \end{bmatrix} \quad \text{and} \quad \det\left(B|AB|A^2B|A^3B\right) = \frac{g^2}{M^4 l^4}$$

Thus, the linearized system is controllable. Therefore, a full state feedback control law $f = -K^T X$ with an appropriate gain vector K is able to successfully stabilize the system to its unstable equilibrium position.

3.5 Stabilizing control law

3.5.1 The homoclinic orbit

Let us first note that in view of (3.12) and (3.5), if $\dot{x} = 0$ and $E(q, \dot{q}) = 0$, then

$$\frac{1}{2} m l^2 \dot{\theta}^2 = mgl(1 - \cos\theta) \qquad (3.16)$$

The above equation defines a very particular trajectory that corresponds to a homoclinic orbit. Note that $\dot{\theta} = 0$ only when $\theta = 0$. This means that the pendulum angular position moves clockwise or counterclockwise until it reaches the equilibrium point $(\theta, \dot{\theta}) = (0,0)$. Thus our objective can be reached if the system can be brought to the orbit (3.16) for $\dot{x} = 0$, $x = 0$ and $E = 0$. Bringing the system to this homoclinic orbit solves the problem of "swinging up" the pendulum. In order to balance the pendulum at the upper equilibrium position, the control must eventually be switched to a controller that guarantees (local) asymptotic stability of this equilibrium [106]. By guaranteeing convergence to the above homoclinic orbit, we guarantee that the trajectory will enter the basin of attraction of any linear balancing controller.

3.5.2 Stabilization around the homoclinic orbit

The passivity property of the system suggests the use of the total energy E in (3.12) in the controller design. Since we wish to bring to zero $x, \dot{x}$ and E, we propose the following Lyapunov function candidate

$$V(q, \dot{q}) = \frac{k_E}{2} E(q, \dot{q})^2 + \frac{k_v}{2} \dot{x}^2 + \frac{k_x}{2} x^2 \qquad (3.17)$$

where k_E, k_v and k_x are strictly positive constants. Note that $V(q, \dot{q})$ is a positive semi-definite function. Differentiating V and using (3.13), we obtain

$$
\begin{aligned}
\dot{V} &= k_E E \dot{E} + k_v \dot{x} \ddot{x} + k_x x \dot{x} \\
&= k_E E \dot{x} f + k_v \dot{x} \ddot{x} + k_x x \dot{x} \\
&= \dot{x}(k_E E f + k_v \ddot{x} + k_x x)
\end{aligned}
\tag{3.18}
$$

Let us now compute $\ddot{x}$ from (3.4). The inverse of $M(q)$ can be obtained from (3.5), (3.6) and (3.8) and is given by

$$
M^{-1} = \frac{1}{\det(M)} \begin{bmatrix} ml^2 & -ml\cos\theta \\ -ml\cos\theta & M+m \end{bmatrix}
\tag{3.19}
$$

with $\det(M) = ml^2(M + m\sin^2\theta)$. Therefore, we have

$$
\begin{bmatrix} \ddot{x} \\ \ddot{\theta} \end{bmatrix} = [\det(M(q))]^{-1} \left(\begin{bmatrix} 0 & m^2 l^3 \dot{\theta} \sin\theta \\ 0 & -m^2 l^2 \dot{\theta} \sin\theta \cos\theta \end{bmatrix} \begin{bmatrix} \dot{x} \\ \dot{\theta} \end{bmatrix} \right.
$$
$$
\left. + \begin{bmatrix} -m^2 l^2 g \sin\theta \cos\theta \\ (M+m)mgl\sin\theta \end{bmatrix} + \begin{bmatrix} ml^2 f \\ -mlf\cos\theta \end{bmatrix} \right)
$$

Thus, $\ddot{x}$ can be written as

$$
\ddot{x} = \frac{1}{M + m\sin^2\theta} \left[m\sin\theta(l\dot{\theta}^2 - g\cos\theta) + f \right]
\tag{3.20}
$$

Introducing the above in (3.18), one has

$$
\dot{V} = \dot{x} \left[f \left(k_E E + \frac{k_v}{M + m\sin^2\theta} \right) + \frac{k_v m \sin\theta(l\dot{\theta}^2 - g\cos\theta)}{M + m\sin^2\theta} + k_x x \right]
\tag{3.21}
$$

We propose a control law such that

$$
f \left(k_E E + \frac{k_v}{M + m\sin^2\theta} \right) + \frac{k_v m \sin\theta(l\dot{\theta}^2 - g\cos\theta)}{M + m\sin^2\theta} + k_x x = -k_\delta \dot{x}
\tag{3.22}
$$

which will lead to

$$\dot{V} = -k_\delta \dot{x}^2 \tag{3.23}$$

Note that other functions $f(\dot{x})$ such that $\dot{x}f(\dot{x}) > 0$ are also possible. The control law in (3.22) will have no singularities, provided that

$$\left(k_E E + \frac{k_v}{M + m\sin^2\theta} \right) \neq 0 \tag{3.24}$$

Note from (3.12) that $E \geq -2mgl$. Thus, (3.24) always holds if the following inequality is satisfied

$$\frac{k_v}{\max_\theta (M + m\sin^2\theta)} > k_E(2mgl) \tag{3.25}$$

This gives the following lower bound for $\frac{k_v}{k_E}$

$$\frac{k_v}{k_E} > 2mgl(M + m) \tag{3.26}$$

Note that when using the control law (3.22), the pendulum can get stuck at the (lower) stable equilibrium point, $(x, \dot{x}, \theta, \dot{\theta}) = (0, 0, \pi, 0)$. In order to avoid this singular point, which occurs when $E = -2mgl$ (see (3.12)), we require

$$|E| < c = 2mgl \tag{3.27}$$

Since V is a non-increasing function (see (3.23)), (3.27) will hold if the initial conditions are such that

$$V(0) < k_E \frac{c^2}{2} \tag{3.28}$$

The above defines the region of attraction as will be shown in the next section.

3.5.3   Domain of attraction

Condition (3.28) imposes bounds on the initial energy of the system. Note that the potential energy $P = mgl(\cos\theta - 1)$ lies between $-2mgl$ and 0. This means that the initial kinetic energy should belong to $[0, c + 2mgl)$.

Note also that the initial position of the cart $x(0)$ is arbitrary, since we can always choose an appropriate value for k_x in V (3.17). If $x(0)$ is large, we should choose a small k_x. The convergence rate of the algorithm may, however, decrease when k_x is small.

Note that when the initial kinetic energy $K(q(0), \dot{q}(0))$ is zero, the initial angular position $\theta(0)$ should belong to $(-\pi, \pi)$. This means that the only forbidden point is $\theta(0) = \pi$. When the initial kinetic energy $K(q(0), \dot{q}(0))$ is different from zero, i.e. $K(q(0), \dot{q}(0))$ belongs to $(0, c + 2mgl)$ (see (3.27) and (3.28)), then there are less restrictions on the initial angular position $\theta(0)$. In particular, $\theta(0)$ can even be pointing downwards, i.e. $\theta = \pi$ provided that $K(q(0), \dot{q}(0))$ is not zero.

Despite the fact that our controller is local, its basin of attraction is far from small. The simulation example and the real-time experiments will show this feature.

For future use we will rewrite the control law f from (3.22) as

$$f = \frac{k_v m \sin\theta \left(g\cos\theta - l\dot{\theta}^2 \right) - (M + m\sin^2\theta)(k_x x + k_\delta \dot{x})}{k_v + (M + m\sin^2\theta) k_E E} \qquad (3.29)$$

3.6   Stability analysis

The stability analysis will be based on LaSalle´s invariance theorem (see for instance [46], page 117).

First, we will reformulate the system, as follows. Since $\cos\theta$ and $\sin\theta$ are bounded functions, we can define z as

$$z = \begin{bmatrix} x \\ \dot{x} \\ \cos\theta \\ \sin\theta \\ \dot{\theta} \end{bmatrix} = \begin{bmatrix} z_1 \\ z_2 \\ z_3 \\ z_4 \\ z_5 \end{bmatrix}$$

The system (3.4)-(3.7) can be written as

$$
\begin{aligned}
\dot{z}_1 &= z_2 \\
\dot{z}_3 &= -z_4 z_5 \\
\dot{z}_4 &= z_3 z_5
\end{aligned}
\tag{3.30}
$$

$$
\begin{bmatrix} \dot{z}_2 \\ \dot{z}_5 \end{bmatrix} = \begin{bmatrix} M+m & mlz_3 \\ mlz_3 & ml^2 \end{bmatrix}^{-1} \left(\begin{bmatrix} f \\ 0 \end{bmatrix} \right.
\tag{3.31}
$$

$$
\left. - \begin{bmatrix} 0 & -mlz_4 z_5 \\ 0 & 0 \end{bmatrix} \begin{bmatrix} z_2 \\ z_5 \end{bmatrix} - \begin{bmatrix} 0 \\ -mglz_4 \end{bmatrix} \right)
$$

The energy E (3.12) is given by

$$
E = \frac{1}{2} \begin{bmatrix} z_2 \\ z_5 \end{bmatrix}^T \begin{bmatrix} M+m & mlz_3 \\ mlz_3 & ml^2 \end{bmatrix} \begin{bmatrix} z_2 \\ z_5 \end{bmatrix} + mgl(z_3 - 1)
\tag{3.32}
$$

The Lyapunov function candidate (3.17) becomes

$$
V = \frac{k_E}{2} E^2 + \frac{k_v}{2} z_2{}^2 + \frac{k_x}{2} z_1{}^2
\tag{3.33}
$$

The derivative of V is then

$$
\dot{V} = z_2(k_E E f + k_v \dot{z}_2 + k_x z_1)
\tag{3.34}
$$

and the control f (3.29) is written as

$$
f = \frac{k_v m z_4 \left(g z_3 - l z_5{}^2\right) - \left(M + m z_4{}^2\right)\left(k_x z_1 + k_\delta z_2\right)}{k_v + \left(M + m z_4{}^2\right) k_E E}
\tag{3.35}
$$

which leads to

$$
\dot{V} = -k_\delta z_2{}^2
\tag{3.36}
$$

Introducing (3.35) into (3.30)-(3.31) we obtain a closed-loop system of the form $\dot{z} = F(z)$. In order to apply LaSalle's theorem, we are required to define a compact (closed and bounded) set Ω with the property that every solution of the system $\dot{z} = F(z)$ that starts in Ω remains in Ω for all future time. Since $V(z_1, z_2, z_3, z_4, z_5)$ in (3.33) is a non-increasing function, (see (3.36)), then z_1, z_2 and z_5 are bounded. Note that z_3 and z_4 are also bounded.

The set Ω is defined as

$$
\Omega = \left\{ z \in \mathbb{R}^5 \,\middle|\, z_3{}^2 + z_4{}^2 = 1, V(z_1, z_2, z_3, z_4, z_5) \le V(z(0)) \right\}
$$

Therefore, the solutions of the closed-loop system $\dot{z} = F(z)$ remain inside a compact set Ω that is defined by the initial value of z. Let Γ be the set of all points in Ω such that $\dot{V}(z) = 0$. Let M be the largest invariant set in Γ. LaSalle's theorem ensures that every solution starting in Ω approaches M as $t \to \infty$. Let us now compute the largest invariant set M in Γ.

In the set Γ (see (3.36)), $\dot{V} = 0$ and $z_2 = 0$, which implies that z_1 and V are constant. From (3.33), it follows that E is also constant. Using (3.31), the expression of $\dot{z}_2$ becomes

$$\dot{z}_2 = \frac{1}{M + mz_4^2} \left[mz_4 \left(lz_5^2 - gz_3 \right) + f \right] \tag{3.37}$$

From (3.37) and (3.35), it follows that the control law has been chosen such that

$$-k_\delta z_2 = k_E Ef + k_v \dot{z}_2 + k_x z_1 \tag{3.38}$$

From the above equation, we conclude that Ef is constant in Γ. Since E is also constant, we either have a) $E = 0$ or b) $E \neq 0$.

- Case a: If $E = 0$, then from (3.38) $z_1 = 0$ (i.e. $x = 0$). Note that f in (3.35) is bounded in view of (3.25)-(3.28). Recall that $E = 0$ means that the trajectories are in the homoclinic orbit (3.16). In this case, we conclude that x, $\dot{x}$, and E converge to zero. Note that if $E = 0$ then f does not necessarily converge to zero.

- Case b: If $E \neq 0$, since Ef is constant, then the control input f is also constant. However, a force input f that is constant and different from zero would lead us to a contradiction. We will give below a mathematical proof of the fact that $f = 0$ in Γ.

Proof 3.1 *We will prove that when $z_2 = 0$, E is constant and $\neq 0$, and f is constant, f should be zero. From (3.31), we get*

$$mlz_5 z_3 - mlz_5^2 z_4 = f \tag{3.39}$$
$$ml^2 \dot{z}_5 - mglz_4 = 0 \tag{3.40}$$

Moreover, the energy E (3.32) is constant and given by

$$E = \frac{1}{2}ml^2 z_5^2 + mgl\,(z_3 - 1) = K_0 \qquad (3.41)$$

Introducing (3.40) in (3.39), we obtain

$$z_4(gz_3 - lz_5{}^2) = \frac{f}{m} \qquad (3.42)$$

The expression (3.41) gives us

$$lz_5^2 = K_1 + 2g\,(1 - z_3)$$

with $K_1 = \frac{2K_0}{ml}$. Combining the above and (3.42)

$$z_4(3gz_3 + K_2) = \frac{f}{m} \qquad (3.43)$$

with $K_2 = -(2g + K_1)$. Taking the time derivative of (3.43), we obtain (see (3.30)-(3.31))

$$z_5\left(3g\left(z_3^2 - z_4^2\right) + K_2 z_3\right) = 0 \qquad (3.44)$$

If $z_5 = 0$, then $\dot{z}_5 = 0$ and from (3.40) we conclude that $z_4 = 0$. If $z_5 \neq 0$, then (3.44) becomes

$$3g\left(z_3^2 - z_4^2\right) + K_2 z_3 = 0 \qquad (3.45)$$

Differentiating (3.45), it follows

$$z_5 z_4\left(-12gz_3 - K_2\right) = 0$$

The case when $z_3 = \frac{-K_2}{12g}$ implies that θ is constant, which implies $z_5 = 0$, and so $z_4 = 0$ (see (3.40)).

In each case we conclude that $z_4 = 0$ and $z_5 = 0$. From (3.39) it follows that $f = 0$. ∎

We therefore conclude that $f = 0$ in Γ. From (3.38), it then follows that $z_1 = 0$ in Γ. It only remains to be proved that $E = 0$ when $z_1 = 0$, $z_2 = 0$ and $f = 0$.

From (3.31), we get

$$ml\dot{z}_5 z_3 - mlz_5{}^2 z_4 = 0 \tag{3.46}$$
$$ml^2 \dot{z}_5 - mglz_4 = 0 \tag{3.47}$$

Introducing (3.47) into (3.46), we obtain

$$\frac{g}{l} z_4 z_3 - z_5{}^2 z_4 = 0 \tag{3.48}$$

Thus, we have either

$$z_5{}^2 = \frac{g}{l} z_3 \tag{3.49}$$

or

$$z_4 = 0 \tag{3.50}$$

Differentiating (3.49), we obtain

$$2 z_5 \dot{z}_5 = -\frac{g}{l} z_5 z_4 \tag{3.51}$$

Let us first study (3.51) and (3.50) afterwards.

- <u>Case 1.</u> If $z_5 \neq 0$, (3.51) becomes

$$2 \dot{z}_5 = -\frac{g}{l} z_4$$

  Combining this equation with (3.47), we conclude that $z_4 = 0$, which implies (3.50).

- <u>Case 2.</u> If $z_5 = 0$ then $\dot{z}_5 = 0$, which together with (3.47) implies that $z_4 = 0$, which implies (3.50).

Also from (3.50) we have $z_4 = 0$, then $\dot{z}_4 = 0$. Since $z_3 = \pm 1$ when $z_4 = 0$, we conclude from (3.30) that $z_5 = 0$. So far we have proved that $z_1 = 0$, $z_2 = 0$, $z_3 = \pm 1$, $z_4 = 0$ and $z_5 = 0$. Moreover, $z_3 = -1$

(which corresponds to $\theta = \pi$ (mod 2π)) has been excluded by imposing condition (3.27) (see also (3.12)). Therefore $z^T = [0, 0, 1, 0, 0]$, which implies that $E = 0$. This contradicts the assumption $E \neq 0$ and thus the only possible case is $E = 0$.

The above discussion can be summarized in the following main result with the original variables $(x, \dot{x}, \theta, \dot{\theta})$.

Lemma 3.1 *Consider the inverted pendulum system (3.4) and the controller in (3.29) with strictly positive constants k_E, k_v, k_x and k_δ satisfying inequality (3.26). Provided that the state initial conditions satisfy inequalities (3.27) and (3.28), then the solution of the closed-loop system converges to the invariant set M given by the homoclinic orbit (3.16) with $(x, \dot{x}) = (0, 0)$. Note that f does not necessarily converge to zero.*
∎

3.7 Simulation results

In order to observe the performance of the proposed control law based on an energy approach of the system, we performed simulations on MATLAB using SIMULINK.

We considered the real system parameters $\bar{M} = M + m = 1.2$, $ml^2 = 0.0097$ and $P = ml = 0.04$, and $g = 9.804 \ ms^{-2}$ of the inverted pendulum at the University of Illinois at Urbana-Champaign. Recall that the control law requires that initial conditions such that (3.28) are satisfied. We chose the gains $k_E = 1$, $k_v = 1$, $k_x = 10^{-2}$ and $k_\delta = 1$. These gains have been chosen to increase the convergence rate in order to switch to a linear stabilizing controller in a reasonable time. Note that these gains satisfy inequality (3.26).

The algorithm brings the inverted pendulum close to the homoclinic orbit but the inverted pendulum will remain swinging while getting closer and closer to the origin. Once the system is close enough to the origin, i.e. ($|x| \leq 0.1, |\dot{x}| \leq 0.2, |\theta| \leq 0.3, |\dot{\theta}| \leq 0.3$), we switch to the linear LQR controller $f = -K[x \ \dot{x} \ \theta \ \dot{\theta}]^T$ where $K = [44 \ \ 23 \ \ 74 \ \ 11]$.

Figures 3.2 and 3.3 show the results for an initial position

$$x = 0.1 \qquad\qquad \dot{x} = 0$$

$$\theta = \frac{2\pi}{3} \qquad\qquad \dot{\theta} = 0$$

Simulations showed that the non-linear control law brings the system to the homoclinic orbit (see the phase plot in Figure 3.3). Switching to the linear controller occurs at time $t = 120$ s. Note that before the switching, the energy E goes to zero and that the Lyapunov function V is decreasing and converges to zero.

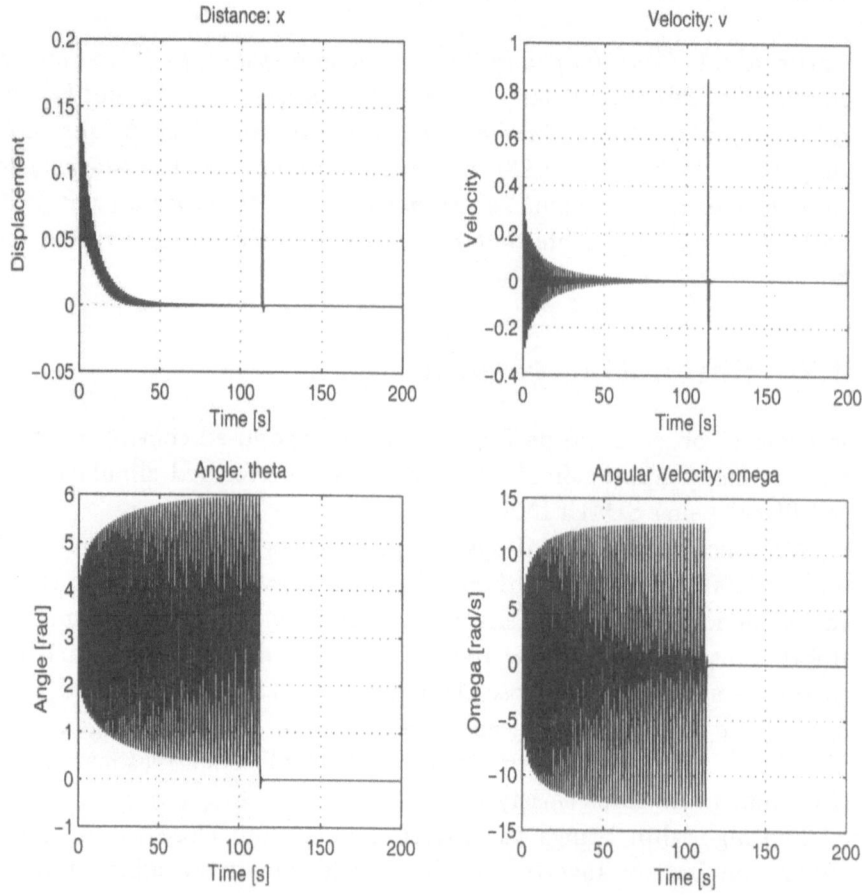

Figure 3.2: Simulation results

3.8   Experimental results

We performed experiments on the inverted pendulum setting at the University of Illinois at Urbana-Champaign. The parameters of the model used for the controller design and the linear controller gains K

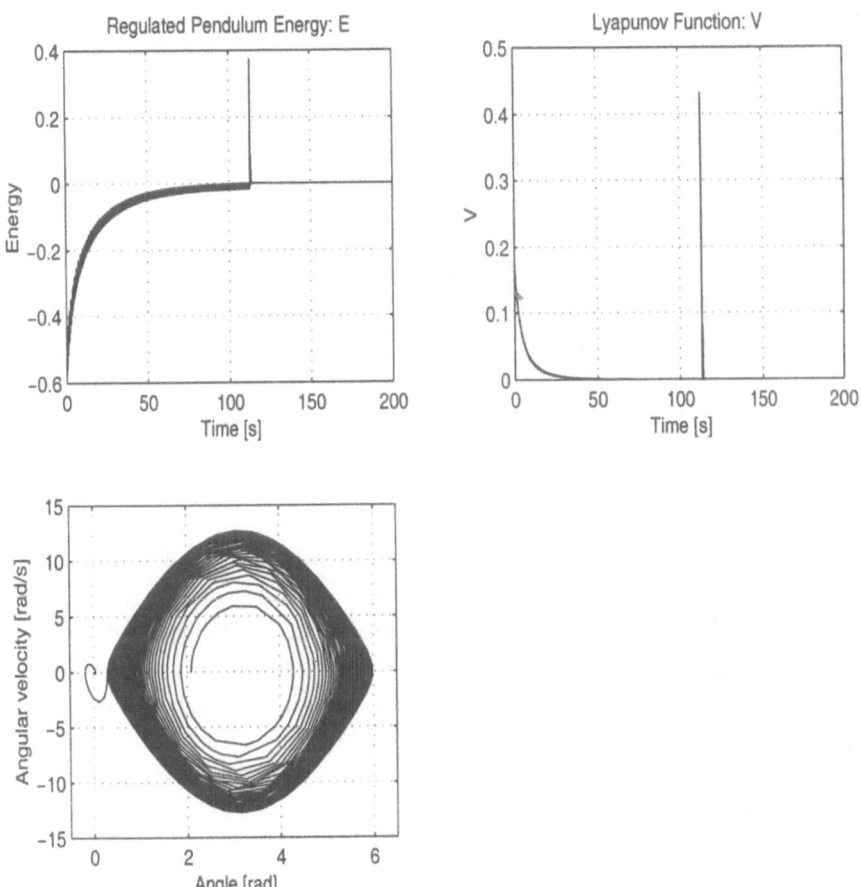

Figure 3.3: Simulation results

are the same as in the previous section.

For this experiment, we chose the gains $k_E = 1$, $k_v = 1.15$, $k_x = 20$ and $k_\delta = 0.001$, which satisfy inequality (3.26).

Figures 3.4 and 3.5 show the results for an initial position

$$x = 0 \qquad\qquad \dot{x} = 0$$

$$\theta = \pi + 0.1 \qquad\qquad \dot{\theta} = 0.1$$

Real-time experiments showed that the non-linear control law brings the system to the homoclinic orbit (see the phase plot in Figure 3.5). Switching to the linear controller occurs at time $t = 27$ s. Note that the control input lies in an acceptable range.

3.9   Conclusions

We have presented a control strategy for the inverted pendulum that brings the pendulum to a homoclinic orbit, while the cart displacement converges to zero. Therefore, the state will enter the basin of attraction of any locally convergent controller.

The control strategy is based on the total energy of the system, using its passivity properties. A Lyapunov function is obtained using the total energy of the system. The convergence analysis is carried out using LaSalle's invariance principle. The system non-linearities have not been compensated before the control design, which has enabled us to exploit the physical properties of the system in the stability analysis.

The control scheme has been tested in a real cart-pole system and good performance has been obtained.

The proposed control strategy is applicable to a wider class of underactuated mechanical systems as we will see in Chapters 4 to 7.

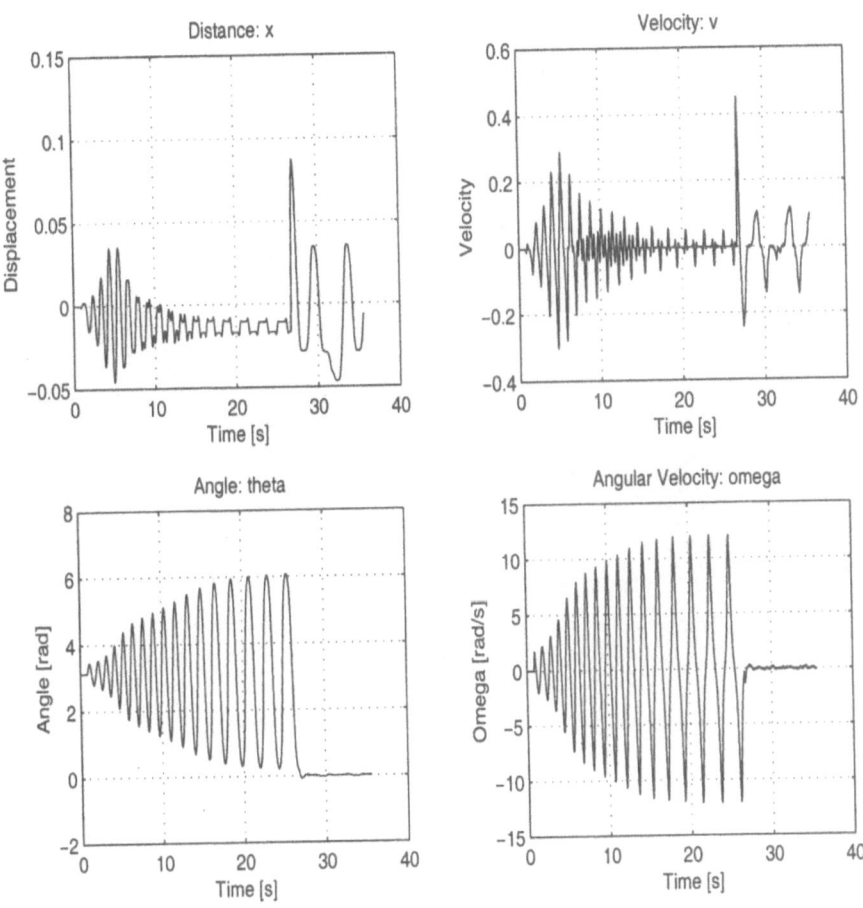

Figure 3.4: Experimental results

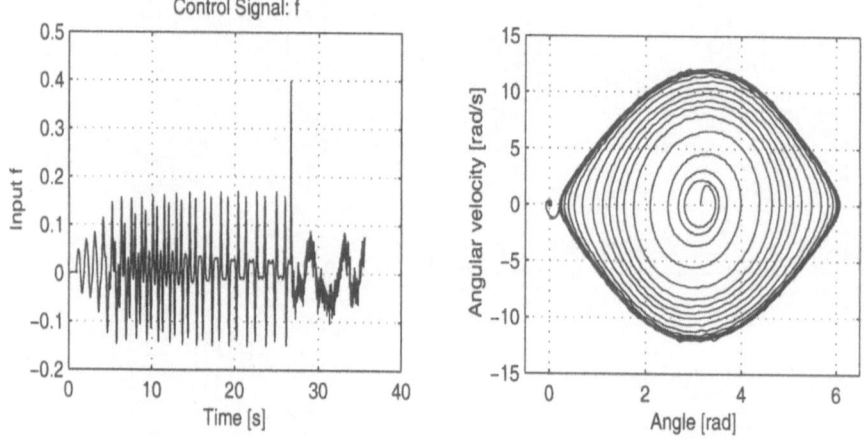

Figure 3.5: Experimental results

Chapter 4

A convey-crane system

4.1 Introduction

Based on the results of control of the inverted pendulum on a cart presented in Chapter 3, we propose a control law of a convey-crane, which transports a load suspended from a cart minimizing the oscillations of the load. The technique has also been presented in [18].

The inverted pendulum on a cart has brought about many contributions, the objective being the stabilization of the unstable equilibrium point as seen in Chapter 3. The problem of asymptotic stabilization of the lower equilibrium point has not been thoroughly studied in the literature. The control objective of a convey-crane presented in this chapter is to move the load to the origin, keeping the oscillations of the suspended mass as small as possible. The system dynamics correspond exactly to the equations of the inverted pendulum on a cart, but now the point of interest is the lower equilibrium point. The stability analysis is carried out using Lyapunov techniques. The present approach takes advantage of the passivity of the model. The non-linearities are not canceled and the control law may be interpreted as adding a non-linear damping to the system dynamics. The performance of the control law is shown in simulations.

4.2 Model

Consider the convey-crane system as shown in Figure 4.1, where M is the mass of the cart, m is the mass of the pendulum and the load of the crane, θ the angle the pendulum makes with the vertical and l the length of the rod. We will assume, as in the case of the inverted pendulum, that the masses are concentrated at their geometrical centers. We will assume that the rod has constant length l and no mass.

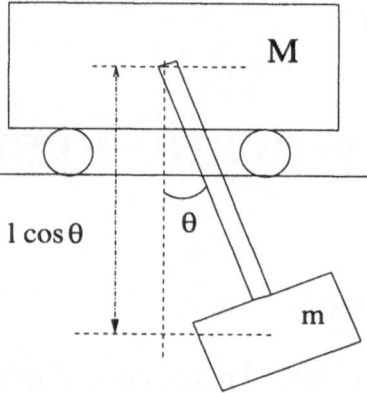

Figure 4.1: The convey-crane system

The equations may be obtained by standard Euler-Lagrange methods or applying Newton's second law. The system dynamics may be described by

$$M(q)\ddot{q} + C(q, \dot{q})\dot{q} + G(q) = \tau \tag{4.1}$$

where

$$q = \begin{bmatrix} x \\ \theta \end{bmatrix} \qquad M(q) = \begin{bmatrix} M + m & -ml\cos\theta \\ -ml\cos\theta & ml^2 \end{bmatrix} \tag{4.2}$$

$$C(q, \dot{q}) = \begin{bmatrix} 0 & ml\sin\theta\,\dot{\theta} \\ 0 & 0 \end{bmatrix} \tag{4.3}$$

$$G(q) = \begin{bmatrix} 0 \\ mgl\sin\theta \end{bmatrix} \text{ and } \quad \tau = \begin{bmatrix} f \\ 0 \end{bmatrix} \tag{4.4}$$

The above model corresponds to the model used for the inverted pendulum on a cart, replacing $\theta \to \theta + \pi$ (see [59] or [46]). Notice that $M(q)$ is symmetric and positive definite, since the parameters M, m, l are positive and

$$
\begin{aligned}
\det\left[M\left(q\right)\right] &= (M+m)\,m\,l^2 - (-m\,l\cos\theta)^2 \\
&= M\,m\,l^2 + m^2 l^2 \sin^2\theta > 0
\end{aligned}
\tag{4.5}
$$

A second well-known property is that the parameters of the model are such that the matrix

$$
\dot{M}(q) - 2C(q,\dot{q}) = \begin{bmatrix} 0 & -m\,l\sin\theta\,\dot{\theta} \\ m\,l\sin\theta\,\dot{\theta} & 0 \end{bmatrix}
\tag{4.6}
$$

is skew-symmetric. This property is required to establish the passivity of the model. Recall [41] that for any skew-symmetric matrix A, $x^T A x = 0$.

Finally, the potential energy associated with the pendulum, may be defined as $P = mgl(1 - \cos\theta)$. With this definition, P and $G(q)$ are related by

$$
G(q) = \frac{\partial P}{\partial q} = \begin{bmatrix} 0 \\ m\,g\,l\sin\theta \end{bmatrix}
\tag{4.7}
$$

4.3 Passivity of the system

The total energy of the system, i.e. the sum of the kinetic energy of the two masses and the potential energy of the pendulum is given by

$$
\begin{aligned}
E &= \tfrac{1}{2}\dot{q}^T M(q)\dot{q} + P(q) \\
&= \tfrac{1}{2}\dot{q}^T M(q)\dot{q} + m\,g\,l\,(1 - \cos\theta)
\end{aligned}
\tag{4.8}
$$

Using (4.1)-(4.4) and (4.6)-(4.7), we may calculate the derivative of the energy E as

$$
\begin{aligned}
\dot{E} &= \dot{q}^T M(q)\ddot{q} + \tfrac{1}{2}\dot{q}^T \dot{M}(q)\dot{q} + \dot{q}^T G(q) \\
&= \dot{q}^T \left(-C\dot{q} - G + \tau + \tfrac{1}{2}\dot{M}(q)\dot{q}\right) + \dot{q}^T G(q) \\
&= \dot{q}^T \tau = \dot{x} f
\end{aligned}
\tag{4.9}
$$

Integrating the last relationship from zero to t, we get

$$
\begin{aligned}
\int_0^t \dot{x} f \, dt' &= E(t) - E(0) \\
&\geq -E(0)
\end{aligned}
\tag{4.10}
$$

which proves the passivity of the system having f as input and $\dot{x}$ as output. Note that when the input force is zero and restricting $\theta \in [0, 2\pi]$, the system (4.1) has two subsets of equilibrium; $\left(x, \dot{x}, \theta, \dot{\theta}\right) = (*, 0, 0, 0)$ is a set of stable equilibrium points and $\left(x, \dot{x}, \theta, \dot{\theta}\right) = (*, 0, \pi, 0)$ corresponds to a set of unstable equilibrium points. The minimum energy corresponds to the lower position of the pendulum and equals zero.

The control objective is to bring the state from the initial conditions $\left(x(0), \dot{x}(0), \theta(0), \dot{\theta}(0)\right) = (x_0, 0, \theta_0, 0)$ to the origin, i.e. change the stable equilibrium point $\left(x, \dot{x}, \theta, \dot{\theta}\right) = (0, 0, 0, 0)$ into an asymptotically stable equilibrium point around some neighborhood of the origin.

4.4   Damping oscillations control law

Notice from (4.8) that if $\dot{x} = 0$ and $E = 0$, then

$$\frac{1}{2} m\, l^2 \dot{\theta}^2 = m\, g\, l \left(\cos \theta - 1\right) \tag{4.11}$$

This is a homoclinic orbit (see Definition 2.10). In order to take advantage of the passivity property of the system, let us propose the following Lyapunov function candidate

$$V(q, \dot{q}) = k_E E(q, \dot{q}) + \frac{k_x}{2} x^2 \tag{4.12}$$

where k_E, k_x are strictly positive constants. The Lyapunov function candidate $V(q, \dot{q})$ is positive definite if we restrict $\theta \in [0, 2\pi)$. Differentiating $V(q, \dot{q})$ and using (4.9), we get

$$\begin{aligned} \dot{V} &= k_E \dot{E} + k_x x \dot{x} \\ &= k_E \dot{x} f + k_x x \dot{x} \\ &= \dot{x} \left(k_E\, f + k_x x\right) \end{aligned} \tag{4.13}$$

From (4.2), we get

$$[M\,(q)]^{-1} = \frac{1}{\det\,(M\,(q))} \left[\begin{array}{cc} m\,l^2 & m\,l\cos\theta \\ m\,l\cos\theta & M + m \end{array} \right] \tag{4.14}$$

with $\det(M(q))$ as in (4.5). From (4.1) and the above, we obtain

$$
\begin{bmatrix} \ddot{x} \\ \ddot{\theta} \end{bmatrix} = \frac{1}{\det(M(q))} \begin{bmatrix} m\,l^2 & m\,l\cos\theta \\ m\,l\cos\theta & M+m \end{bmatrix} \tag{4.15}
$$

$$
\left\{ -\begin{bmatrix} m\,l\sin\theta\,\dot{\theta}^2 \\ 0 \end{bmatrix} - \begin{bmatrix} 0 \\ m\,g\,l\sin\theta \end{bmatrix} + \begin{bmatrix} f \\ 0 \end{bmatrix} \right\} \tag{4.16}
$$

From the above, we get

$$
\ddot{x} = \frac{1}{M+m\sin^2\theta} \left[-m\sin\theta \left(l\,\dot{\theta}^2 + g\cos\theta \right) + f \right] \tag{4.17}
$$

and

$$
\ddot{\theta} = \frac{1}{l(M+m\sin^2\theta)} \left[-\sin\theta\left((M+m)g + m\,l\,\cos\theta\,\dot{\theta}^2\right) + \cos\theta f \right] \tag{4.18}
$$

For the sake of simplicity, we will consider $M = m = l = 1$, then we will propose the control law such that

$$
(k_E\,f + k_x x) = -\gamma\dot{x} \tag{4.19}
$$

for some $\gamma > 0$, which leads to

$$
\dot{V} = -\gamma\dot{x}^2 \tag{4.20}
$$

The explicit control law defined by (4.19) is

$$
f = -\frac{1}{k_E}(k_x x + \gamma\dot{x}) \tag{4.21}
$$

The control law (4.21) guarantees $\dot{V} = -\gamma\dot{x}^2$, which is negative semi-definite, therefore the closed loop is stable [46].

4.4.1 Asymptotic stability analysis

Using LaSalle's invariance principle, we will prove the following

Theorem 4.1 *The closed-loop system given by equations (4.1) and (4.21) is such that the origin is asymptotically stable for all points in $I\!R^4 \setminus \{(0,0,\pi,0)\}$.* ■

Proof 4.1 *From (4.20), the invariance set Γ is $\Gamma : \{(x,\dot{x},\theta,\dot{\theta}) : \dot{x} = 0\}$. Thus, x is constant in Γ. Assume that in Γ, $x = a \neq 0$. Then, from (4.21), $f = -\frac{k_x}{k_E}a \neq 0$, which leads to a contradiction since a constant force will eventually produce a displacement of the cart. Therefore, in Γ, $x = 0$ and $f = 0$. Since the cart is at rest, the pendulum can be either at rest or oscillating. However, an oscillatory movement of the pendulum would exert a force on the cart and produce a displacement. Since $x = 0$ in Γ, we conclude that $\theta = 0$ or $\theta = n\pi$ in Γ. This can also be concluded from (4.17), which for $x = 0$ reduces to*

$$\sin\theta\left(l\dot{\theta}^2 + g\cos\theta\right) = 0 \qquad (4.22)$$

Note that the equation $\dot{\theta}^2 = -\frac{g}{l}\cos\theta$ has an equilibrium at $\theta = \pm\frac{\pi}{2}$ that is unrealistic, since $f = 0$ in Γ. Therefore, the only possible solution of (4.22) is $\theta = 0$ or $\theta = n\pi$, $n = 1,2,...$ We either have convergence to $(x,\dot{x},\theta,\dot{\theta}) = (0,0,0[2\pi],0)$ for which $E = 0$ or to $(x,\dot{x},\theta,\dot{\theta}) = (0,0,\pi[2\pi],0)$ for which $E = 2mgl$. The latter convergence point can be avoided by constraining the initial conditions to the region $V(0) < 2k_E mgl$. ■

4.5 Simulation results

For comparison reasons, we obtained from (4.17) and (4.18) a linearized model of the convey-crane around its lower equilibrium point and with a force $f = 0$.

$$\dot{z} = \begin{bmatrix} 0 & 1 & 0 & 0 \\ 0 & 0 & -\frac{mg}{M} & 0 \\ 0 & 0 & 0 & 1 \\ 0 & 0 & -\frac{(M+m)g}{lM} & 0 \end{bmatrix} z + \begin{bmatrix} 0 \\ \frac{1}{M} \\ 0 \\ \frac{1}{lM} \end{bmatrix} f \qquad (4.23)$$

We used full state feedback $f = -kz$ on the linearized model with $k = \begin{bmatrix} 3 & 3.69 & 0.71 & -0.87 \end{bmatrix}$. Simulations were performed using SIMULINK, we considered $M = 1$, $m = 1$, $l = 1$ and $g = 9.8$ m/s^2, and the initial conditions were $\left(x(0), \dot{x}(0), \theta(0), \dot{\theta}(0)\right) = (-5, 0, -\pi/4, 0)$.

The parameters of the control law (4.21) were $k_E = 1$, $k_x = 3$ and $\gamma = 4.3$. Figure 4.2 shows the position of the cart x and the angle θ for the original and linearized models with their respective controllers. Figure 4.3 shows the angle θ and the position of the cart x, now for the initial conditions $(-5, 0, 0, 0)$. In these two cases and for different initial conditions, the proposed controller outperforms the linearized controller.

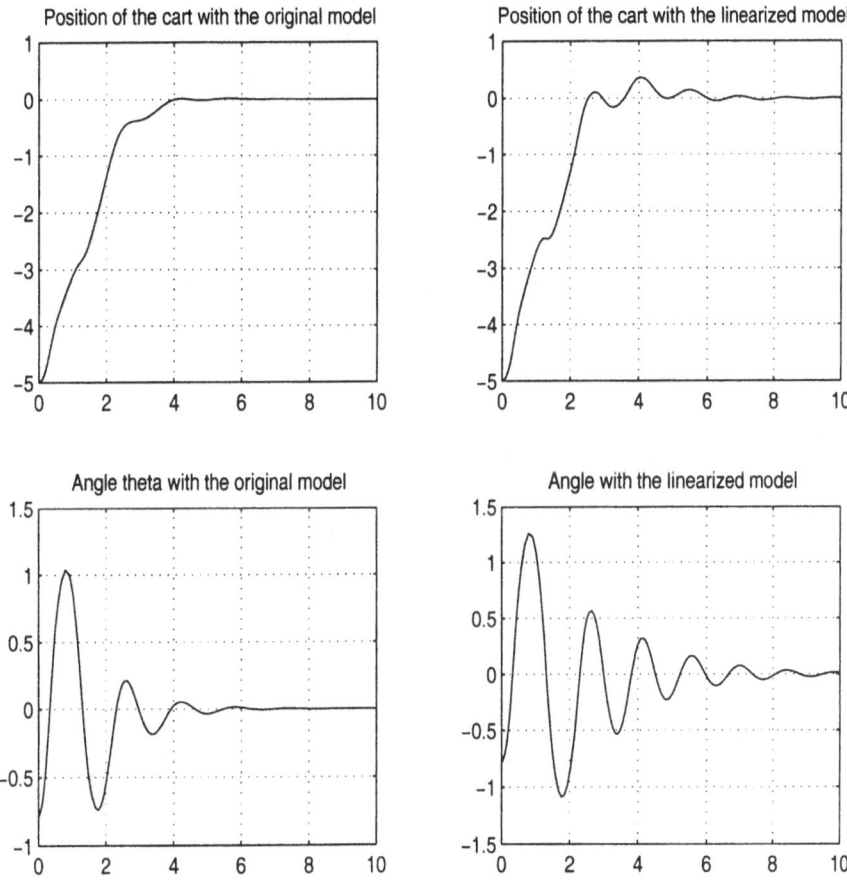

Figure 4.2: Cart position and angle θ for an initial position $(-5, 0, -\frac{\pi}{4}, 0)$

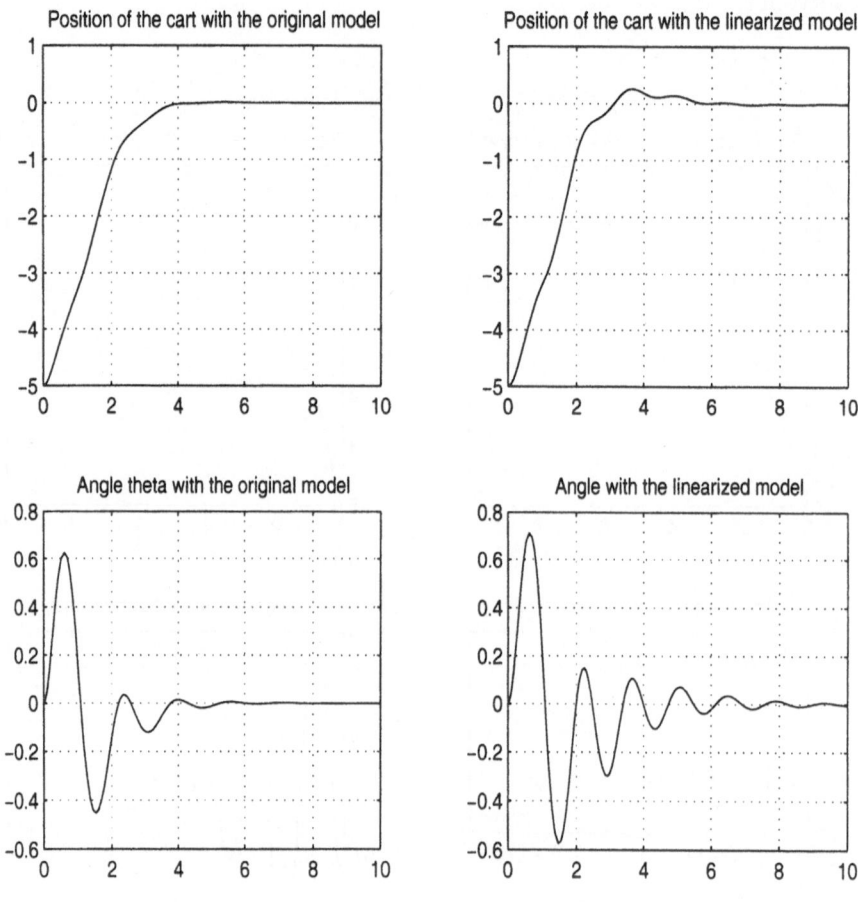

Figure 4.3: Cart position and angle θ for an initial position $(-5, 0, 0, 0)$

4.6 Concluding remarks

We have presented a control law for the convey-crane model, which is similar to the inverted pendulum on a cart, considering the lower equilibrium point as the control objective. We proved asymptotic stability of the proposed control law using a Lyapunov function, which is based on the energy of the system. The convergence analysis was completed using LaSalle's invariance theorem. Simulations show that the region of attraction is practically the whole state space.

Chapter 5

The pendubot system

5.1 Introduction

The two-link underactuated robotic mechanism called the pendubot is used for research in non-linear control and for education in various concepts like non-linear dynamics, robotics and control system design.

This device is a two-link planar robot with an actuator at the shoulder (link 1) and no actuator at the elbow (link 2). The link 2 moves freely around link 1 and the control objective is to bring the mechanism to the unstable equilibrium points.

Similar mechanical systems are numerous: the single and double inverted pendulum, the acrobot [11], the underactuated planar robot [1], etc. Control strategies for the inverted pendulum have been proposed in [4, 59, 86, 98].

Block [10] proposed a control strategy based on two control algorithms to control the pendubot. For the swing up control, Spong and Block [107] used partial feedback linearization techniques and for the balancing and stabilizing controller, they used linearization about the desired equilibrium point by Linear Quadratic Regulator (LQR) and pole placement techniques. The upright position is reached quickly as shown by an application. Nevertheless, they do not present a stability analysis. The authors used concepts such as partial feedback linearization, zero dynamics, and relative degree and discussed the use of the pendubot for educational purposes. To our knowledge, there exists only this solution in the literature to solve the swing up problem of the pendubot.

The controller that we propose is not based on the standard techniques of feedback linearization (or partial feedback linearization). We

53

believe that our approach is the first for which a complete stability analysis has been presented.

The stabilization algorithm proposed here is an adaptation of the work of [24] and Chapter 3, which deals with the inverted pendulum. We will consider the passivity properties of the pendubot and use an energy-based approach to establish the proposed control law. The control algorithm as well as the convergence analysis are based on Lyapunov theory.

This chapter is organized as follows. In Section 5.2, the model of the pendubot system is presented. Section 5.3 gives the passivity properties of the system. In Section 5.4, the controllability of the linearized system is studied. In Section 5.5, the stabilization of the system around its homoclinic orbit using an energy approach is proposed. In Section 5.6, we present the stability analysis of the proposed control law. The performance of the control law is shown in a simulation example and in real-time experiments, in Sections 5.7 and 5.8. We conclude this chapter with some final remarks in Section 5.9.

5.2 System dynamics

Consider the two-link underactuated planar robot, called the pendubot. We will consider the standard assumption, i.e. no friction, etc..

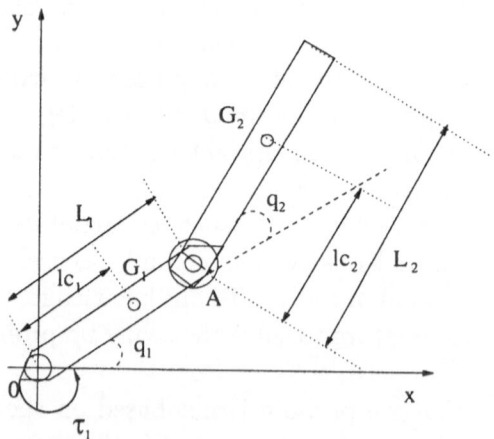

Figure 5.1: The pendubot system

m_1	:	Mass of link 1
m_2	:	Mass of link 2
l_1	:	Length of link 1
l_2	:	Length of link 2
l_{c_1}	:	Distance to the center of mass of link 1
l_{c_2}	:	Distance to the center of mass of link 2
I_1	:	Moment of inertia of link 1 about its centroid
I_2	:	Moment of inertia of link 2 about its centroid
g	:	Acceleration due to gravity
q_1	:	Angle that link 1 makes with the horizontal
q_2	:	Angle that link 2 makes with link 1
τ_1	:	Torque applied on link 1

We have introduced the following five parameter equations

$$\left\{ \begin{array}{ll} \theta_1 & = m_1 l_{c_1}^2 + m_2 l_1^2 + I_1 \\ \theta_2 & = m_2 l_{c_2}^2 + I_2 \\ \theta_3 & = m_2 l_1 l_{c_2} \\ \theta_4 & = m_1 l_{c_1} + m_2 l_1 \\ \theta_5 & = m_2 l_{c_2} \end{array} \right. \tag{5.1}$$

For a control design that neglects friction, these five parameters are all that are needed.

5.2.1 Equations of motion via Euler-Lagrange formulation

We first present the kinetic and potential energies that are used to compute the Lagrangian function. The kinetic energy of link 1 is

$$K_1 = \frac{1}{2} \left(I_1 + m_1 l_{c_1}^2 \right) \dot{q}_1^2$$

The kinetic energy of link 2 is

$$\begin{aligned} K_2 & = \frac{1}{2} \left(I_2 + m_2 l_1 l_{c_2} \cos q_2 + m_2 l_{c_2}^2 + m_2 l_1^2 \right) \dot{q}_1^2 \\ & + \left(I_2 + m_2 l_1 l_{c_2} \cos q_2 + m_2 l_{c_2}^2 \right) \dot{q}_1 \dot{q}_2 + \frac{1}{2} \left(I_2 + m_2 l_{c_2}^2 \right) \dot{q}_2^2 \end{aligned}$$

With the five parameters defined in (5.1), the total kinetic energy is

$$\begin{aligned} K & = K_1 + K_2 \\ K & = \frac{1}{2} \left(\theta_1 + \theta_2 + 2\theta_3 \cos q_2 \right) \dot{q}_1^2 + \frac{1}{2}\theta_2 \dot{q}_2^2 + \left(\theta_2 + \theta_3 \cos q_2 \right) \dot{q}_1 \dot{q}_2 \end{aligned}$$

The total potential energy is $P = \theta_4 g \sin q_1 + \theta_5 g \sin(q_1 + q_2)$. The Lagrangian function is given by

$$
\begin{aligned}
L &= K - P \\
L &= \frac{1}{2}\left(\theta_1 + \theta_2 + 2\theta_3 \cos q_2\right)\dot{q}_1^2 + \frac{1}{2}\theta_2 \dot{q}_2^2 + \left(\theta_2 + \theta_3 \cos q_2\right)\dot{q}_1 \dot{q}_2 \\
&\quad -\theta_4 g \sin q_1 - \theta_5 g \sin(q_1 + q_2)
\end{aligned}
$$

The corresponding equations of motion are derived using Lagrange's equations

$$
\frac{d}{dt}\left(\frac{\partial L}{\partial \dot{q}}(q,\dot{q})\right) - \frac{\partial L}{\partial q}(q,\dot{q}) = \tau \tag{5.2}
$$

where $q = (q_1, ...q_n)^T$ represents the generalized variables, one for each degree of freedom of the system, $\tau = (\tau_1, ..., \tau_n)^T$ denotes forces that are externally applied to the system. In our case, the generalized variables are q_1 and q_2, i.e. $q = (q_1, q_2)^T$ and $\tau = (\tau_1, 0)^T$, where τ_1 is the force applied on the first link. We therefore have

$$
\begin{aligned}
\left(\frac{\partial L}{\partial \dot{q}_1}\right) &= \left(\theta_1 + \theta_2 + 2\theta_3 \cos q_2\right)\dot{q}_1 + \left(\theta_2 + \theta_3 \cos q_2\right)\dot{q}_2 \\
\left(\frac{\partial L}{\partial q_1}\right) &= -\theta_4 g \cos q_1 - \theta_5 g \cos(q_1 + q_2) \\
\left(\frac{\partial L}{\partial \dot{q}_2}\right) &= \theta_2 \dot{q}_2 + \left(\theta_2 + \theta_3 \cos q_2\right)\dot{q}_1 \\
\left(\frac{\partial L}{\partial q_2}\right) &= -\theta_3 \sin q_2 \dot{q}_1^2 - \theta_3 \sin(q_2)\dot{q}_1 \dot{q}_2 - \theta_5 g \cos(q_1 + q_2)
\end{aligned} \tag{5.3}
$$

From Lagrange's equations (5.3), we finally obtain the equations of motion (5.4) and (5.5)

$$
\begin{aligned}
\tau_1 &= \left(\theta_1 + \theta_2 + 2\theta_3 \cos q_2\right)\ddot{q}_1 + \left(\theta_2 + \theta_3 \cos q_2\right)\ddot{q}_2 - \theta_3 \sin q_2 \dot{q}_2^2 \\
&\quad -2\theta_3 \sin q_2 \dot{q}_1 \dot{q}_2 + \theta_4 g \cos q_1 + \theta_5 g \cos(q_1 + q_2) \tag{5.4}
\end{aligned}
$$

$$
\begin{aligned}
0 &= \theta_2 \ddot{q}_2 + \left(\theta_2 + \theta_3 \cos q_2\right)\ddot{q}_1 + \theta_3 \sin q_2 \dot{q}_1^2 \\
&\quad +\theta_5 g \cos(q_1 + q_2) \tag{5.5}
\end{aligned}
$$

They can be rewritten in standard form (5.6)

$$D(q)\ddot{q} + C(q,\dot{q})\dot{q} + g(q) = \tau \tag{5.6}$$

where

$$q = \begin{bmatrix} q_1 \\ q_2 \end{bmatrix} \qquad D(q) = \begin{bmatrix} \theta_1 + \theta_2 + 2\theta_3 \cos q_2 & \theta_2 + \theta_3 \cos q_2 \\ \theta_2 + \theta_3 \cos q_2 & \theta_2 \end{bmatrix} \tag{5.7}$$

$$C(q,\dot{q}) = \begin{bmatrix} -\theta_3 \sin(q_2)\,\dot{q}_2 & -\theta_3 \sin(q_2)\,\dot{q}_2 - \theta_3 \sin(q_2)\,\dot{q}_1 \\ \theta_3 \sin(q_2)\,\dot{q}_1 & 0 \end{bmatrix} \tag{5.8}$$

$$g(q) = \begin{bmatrix} \theta_4 g \cos q_1 + \theta_5 g \cos(q_1+q_2) \\ \theta_5 g \cos(q_1+q_2) \end{bmatrix} \quad \text{and} \quad \tau = \begin{bmatrix} \tau_1 \\ 0 \end{bmatrix} \tag{5.9}$$

Note that $D(q)$ is symmetric. Moreover,

$$
\begin{aligned}
d_{11} &= \theta_1 + \theta_2 + 2\theta_3 \cos q_2 \\
&= m_1 l_{c_1}^2 + m_2 l_1^2 + I_1 + m_2 l_{c_2}^2 + I_2 + 2m_2 l_1 l_{c_2} \cos q_2 \\
&\geq m_1 l_{c_1}^2 + m_2 l_1^2 + I_1 + m_2 l_{c_2}^2 + I_2 - 2m_2 l_1 l_{c_2} \\
&\geq m_1 l_{c_1}^2 + I_1 + I_2 + m_2 (l_1 - l_{c_2})^2 > 0
\end{aligned}
$$

and

$$
\begin{aligned}
\det(D(q)) &= \theta_1 \theta_2 - \theta_3^2 \cos^2 q_2 \\
&= \left(m_1 l_{c_1}^2 + I_1\right)\left(m_2 l_{c_2}^2 + I_2\right) + m_2 l_1^2 I_2 + m_2^2 l_1^2 l_{c_2}^2 \sin^2 q_2 \\
&> 0
\end{aligned}
$$

Therefore $D(q)$ is positive definite for all q. From (5.8), it follows that

$$\dot{D}(q) - 2C(q,\dot{q}) = \begin{bmatrix} 0 & \theta_3 \sin q_2\,(2\dot{q}_1 + \dot{q}_2) \\ -\theta_3 \sin q_2\,(2\dot{q}_1 + \dot{q}_2) & 0 \end{bmatrix} \tag{5.10}$$

which is a skew-symmetric matrix. An important property of skew-symmetric matrices, which will be used in establishing the passivity property of the pendubot, is

$$z^T(\dot{D}(q) - 2C(q, \dot{q}))z = 0 \qquad \forall z \qquad\qquad (5.11)$$

The potential energy of the pendubot can be defined as $P(q) = \theta_4 g \sin q_1 + \theta_5 g \sin(q_1 + q_2)$. Note that P is related to $g(q)$ as follows

$$g(q) = \frac{\partial P}{\partial q} = \begin{bmatrix} \theta_4 g \cos q_1 + \theta_5 g \cos(q_1 + q_2) \\ \theta_5 g \cos(q_1 + q_2) \end{bmatrix} \qquad (5.12)$$

5.3 Passivity of the pendubot

The total energy of the pendubot is given by

$$
\begin{aligned}
E &= \tfrac{1}{2}\dot{q}^T D(q)\dot{q} + P(q) \\
&= \tfrac{1}{2}\dot{q}^T D(q)\dot{q} + \theta_4 g \sin q_1 + \theta_5 g \sin(q_1 + q_2)
\end{aligned}
\qquad (5.13)
$$

Therefore, from (5.6), (5.7), (5.9) and (5.11), we obtain

$$
\begin{aligned}
\dot{E} &= \dot{q}^T D(q)\ddot{q} + \tfrac{1}{2}\dot{q}^T \dot{D}(q)\dot{q} + \dot{q}^T g(q) \\
&= \dot{q}^T(-C(q, \dot{q})\dot{q} - g(q) + \tau) + \tfrac{1}{2}\dot{q}^T \dot{D}(q)\dot{q} + \dot{q}^T g(q) \qquad (5.14) \\
&= \dot{q}^T \tau = \dot{q}_1 \tau_1
\end{aligned}
$$

Integrating both sides of the above equation, we obtain

$$\int_0^t \dot{q}_1 \tau_1 dt = E(t) - E(0) \qquad\qquad (5.15)$$

Therefore, the system having τ_1 as input and $\dot{q}_1$ as output is passive. Note that for $\tau_1 = 0$, the system (5.6) has four equilibrium points; $(q_1, \dot{q}_1, q_2, \dot{q}_2) = (\frac{\pi}{2}, 0, 0, 0)$ and $(q_1, \dot{q}_1, q_2, \dot{q}_2) = (-\frac{\pi}{2}, 0, \pi, 0)$ are two unstable equilibrium positions (respectively, top position and mid position). We wish to reach the top position. $(q_1, \dot{q}_1, q_2, \dot{q}_2) = (\frac{\pi}{2}, 0, \pi, 0)$ is an unstable equilibrium position that we want to avoid, and $(q_1, \dot{q}_1, q_2, \dot{q}_2) = (-\frac{\pi}{2}, 0, 0, 0)$ is the stable equilibrium position we want to avoid too. The total energy $E(q, \dot{q})$ is different for each of the four equilibrium positions

$$
\begin{aligned}
E\left(\tfrac{\pi}{2},0,0,0\right) &= E_{top} = (\theta_4 + \theta_5)\,g && \text{Top positions for both links} \\
E\left(-\tfrac{\pi}{2},0,0,0\right) &= E_{l_1} = (-\theta_4 - \theta_5)\,g && \text{Low positions for both links} \\
E\left(-\tfrac{\pi}{2},0,\pi,0\right) &= E_{mid} = (\theta_5 - \theta_4)g && \text{Mid position: low for link 1} \\
&&& \text{and up for link 2} \\
E\left(\tfrac{\pi}{2},0,\pi,0\right) &= E_{l_2} = (\theta_4 - \theta_5)g && \text{Position: up for link 1} \\
&&& \text{and low for link 2}
\end{aligned}
$$

$$\tag{5.16}$$

The control objective is to stabilize the system around its top unstable equilibrium position.

5.4 Linearization of the system

In the same manner as for the inverted pendulum (3.4), we will linearize the pendubot's non-linear equations of motion (5.4) and (5.5) about the top equilibrium position. Let us first rewrite (5.4) and (5.5) as follows

$$
\begin{aligned}
\ddot{q}_1 = \frac{1}{\theta_1\theta_2 - \theta_3^2\cos^2 q_2} \Big[&\theta_2\theta_3\sin q_2\,(\dot q_1 + \dot q_2)^2 + \theta_3^2\cos q_2\sin(q_2)\,\dot q_1^2 \\
&-\theta_2\theta_4 g\cos q_1 + \theta_3\theta_5 g\cos q_2\cos(q_1 + q_2) + \theta_2\tau_1 \Big]
\end{aligned}
$$

$$\tag{5.17}$$

$$
\begin{aligned}
\ddot{q}_2 = \frac{1}{\theta_1\theta_2 - \theta_3^2\cos^2 q_2} \Big[&-\theta_3(\theta_2 + \theta_3\cos q_2)\sin q_2\,(\dot q_1 + \dot q_2)^2 \\
&-(\theta_1 + \theta_3\cos q_2)\theta_3\sin(q_2)\,\dot q_1^2 + (\theta_2 + \theta_3\cos q_2)(\theta_4 g\cos q_1 - \tau_1) \\
&-(\theta_1 + \theta_3\cos q_2)\theta_5 g\cos(q_1 + q_2) \Big]
\end{aligned}
$$

$$\tag{5.18}$$

Consider the system state $Y = [q_1, \dot q_1, q_2, \dot q_2]$. Differentiating equations (5.17) and (5.18) with respect to the states and evaluating them at the top unstable equilibrium position leads to the following linear system

$$
\frac{d}{dt}
\begin{bmatrix} q_1 \\ \dot q_1 \\ q_2 \\ \dot q_2 \end{bmatrix}
=
\begin{bmatrix}
0 & 1 & 0 & 0 \\
\frac{(\theta_2\theta_4 - \theta_3\theta_5)g}{\theta_1\theta_2 - \theta_3^2} & 0 & -\frac{\theta_3\theta_5 g}{\theta_1\theta_2 - \theta_3^2} & 0 \\
0 & 0 & 0 & 1 \\
\frac{\theta_5 g(\theta_1 + \theta_3) - \theta_4 g(\theta_2 + \theta_3)}{\theta_1\theta_2 - \theta_3^2} & 0 & \frac{\theta_5 g(\theta_1 + \theta_3)}{\theta_1\theta_2 - \theta_3^2} & 0
\end{bmatrix}
\begin{bmatrix} q_1 \\ \dot q_1 \\ q_2 \\ \dot q_2 \end{bmatrix}
$$

$$
+
\begin{bmatrix}
0 \\ \frac{\theta_2}{\theta_1\theta_2 - \theta_3^2} \\ 0 \\ \frac{-\theta_2 - \theta_3}{\theta_1\theta_2 - \theta_3^2}
\end{bmatrix}
\tau_1 = AY + B\tau_1
$$

We then have

$$B = \begin{bmatrix} 0 \\ \frac{\theta_2}{\theta_1\theta_2-\theta_3^2} \\ 0 \\ \frac{-\theta_2-\theta_3}{\theta_1\theta_2-\theta_3^2} \end{bmatrix} \qquad AB = \begin{bmatrix} \frac{\theta_2}{\theta_1\theta_2-\theta_3^2} \\ 0 \\ \frac{-\theta_2-\theta_3}{\theta_1\theta_2-\theta_3^2} \\ 0 \end{bmatrix}$$

$$A^2B = \begin{bmatrix} 0 \\ \frac{g(\theta_4\theta_2^2+\theta_5\theta_3^2)}{\theta_1\theta_2-\theta_3^2} \\ 0 \\ -\frac{g(\theta_4\theta_2^2+\theta_2\theta_4\theta_3+\theta_5\theta_1\theta_3+\theta_5\theta_3^2)}{\theta_1\theta_2-\theta_3^2} \end{bmatrix}$$

$$A^3B = \begin{bmatrix} \frac{g(\theta_4\theta_2^2+\theta_5\theta_3^2)}{\theta_1\theta_2-\theta_3^2} \\ 0 \\ -\frac{g(\theta_4\theta_2^2+\theta_2\theta_4\theta_3+\theta_5\theta_1\theta_3+\theta_5\theta_3^2)}{\theta_1\theta_2-\theta_3^2} \\ 0 \end{bmatrix}$$

and $\det\left(B|AB|A^2B|A^3B\right) = \frac{g^2\theta_5^2\theta_3^2}{(\theta_1\theta_2-\theta_3^2)^4}$. Thus, the linearized system is controllable. Therefore, a full state feedback controller $\tau_1 = -K^TY$ with an appropriate gain vector K is able to successfully stabilize the system to its top unstable equilibrium position.

5.5   Control law for the top position

5.5.1   The homoclinic orbit

Let us first note that in view of (5.13), (5.7), and (5.8), if the following conditions are satisfied

$$\begin{aligned} c_1) \quad & \dot{q}_1 = 0 \\ c_2) \quad & E\left(q,\dot{q}\right) = \left(\theta_4+\theta_5\right)g \end{aligned}$$

then

$$E\left(q,\dot{q}\right) = \frac{1}{2}\theta_2\dot{q}_2^2 + \theta_4 g \sin q_1 + \theta_5 g \sin\left(q_1+q_2\right) = \theta_4 g + \theta_5 g \qquad (5.19)$$

From the above, it follows that if $q_1 \neq \frac{\pi}{2}$ then $\dot{q}_2^2 > 0$. If in addition to conditions c_1) and c_2) we also have condition c_3) $q_1 = \frac{\pi}{2}$, then (5.19) gives

$$\frac{1}{2}\theta_2\dot{q}_2^2 = \theta_5 g \left(1 - \cos q_2\right) \tag{5.20}$$

The above equation defines a very particular trajectory that corresponds to a homoclinic orbit. This means that the link 2 angular position moves clockwise or counter-clockwise until it reaches the equilibrium point $(q_2, \dot{q}_2) = (0, 0)$. Thus, our objective can be reached if the system can be brought to the orbit (5.20) for $\dot{q}_1 = 0$ and $q_1 = \frac{\pi}{2}$. Bringing the system to this homoclinic orbit solves the "swing up" problem. In order to balance the pendubot at the top equilibrium configuration $(\pi/2, 0, 0, 0)$, the control must eventually be switched to a controller that guarantees (local) asymptotic stability of this equilibrium. Such a balancing controller can be designed using several methods, for example LQR, which in fact provides local exponential stability of the top equilibrium. By guaranteeing convergence to the above homoclinic orbit, we guarantee that the trajectory will eventually enter the basin of attraction of any balancing controller.

5.5.2 Stabilization around the homoclinic orbit

The passivity property of the system suggests us to use the total energy E in (5.13) in the controller design. Let us consider $\tilde{q}_1 = \left(q_1 - \frac{\pi}{2}\right)$ and $\tilde{E} = (E - E_{top})$. We wish to bring to zero $\tilde{q}_1$, $\dot{q}_1$ and $\tilde{E}$. We propose the following Lyapunov function candidate

$$V(q, \dot{q}) = \frac{k_E}{2}\tilde{E}(q, \dot{q})^2 + \frac{k_D}{2}\dot{q}_1^2 + \frac{k_P}{2}\tilde{q}_1^2 \tag{5.21}$$

where k_E, k_D and k_P are strictly positive constants to be defined later. Note that $V(q, \dot{q})$ is a positive semi-definite function. Differentiating V and using (5.14), we obtain

$$\begin{aligned}\dot{V} &= k_E\tilde{E}\dot{E} + k_D\dot{q}_1\ddot{q}_1 + k_P\tilde{q}_1\dot{q}_1 \\ &= k_E\tilde{E}\dot{q}_1\tau_1 + k_D\dot{q}_1\ddot{q}_1 + k_P\tilde{q}_1\dot{q}_1 \\ &= \dot{q}_1(k_E\tilde{E}\tau_1 + k_D\ddot{q}_1 + k_P\tilde{q}_1)\end{aligned} \tag{5.22}$$

Let us now compute $\ddot{q}_1$ from (5.6). The inverse of $D(q)$ can be obtained from (5.7) and (5.10) and is given by

$$D^{-1}(q) = [\det(D(q))]^{-1} \begin{bmatrix} \theta_2 & -\theta_2 - \theta_3 \cos q_2 \\ -\theta_2 - \theta_3 \cos q_2 & \theta_1 + \theta_2 + 2\theta_3 \cos q_2 \end{bmatrix}$$

$$(5.23)$$

with

$$\det(D(q)) = \theta_1 \theta_2 - \theta_3^2 \cos^2 q_2$$

Therefore, we have

$$\begin{bmatrix} \ddot{q}_1 \\ \ddot{q}_2 \end{bmatrix} = [\det(D(q))]^{-1} \left(\begin{array}{c} \theta_2 \tau_1 \\ -(\theta_2 + \theta_3 \cos q_2)\, \tau_1 \end{array} \right)$$
$$-D^{-1}(q) \left(C(q, \dot{q}) \begin{bmatrix} \dot{q}_1 \\ \dot{q}_2 \end{bmatrix} + g(q) \right)$$

$\ddot{q}_1$ can thus be written as

$$\ddot{q}_1 = \frac{1}{\theta_1 \theta_2 - \theta_3^2 \cos^2 q_2} \left[\theta_2 \tau_1 + \theta_2 \theta_3 \sin q_2 \left(\dot{q}_1 + \dot{q}_2 \right)^2 \right.$$
$$\left. \theta_3^2 \cos q_2 \sin(q_2)\, \dot{q}_1^2 - \theta_2 \theta_4 g \cos q_1 + \theta_3 \theta_5 g \cos q_2 \cos(q_1 + q_2) \right]$$

To reduce the expressions, we will consider

$$F(q_1, \dot{q}_1, q_2, \dot{q}_2) = \theta_2 \theta_3 \sin q_2 \left(\dot{q}_1 + \dot{q}_2 \right)^2 + \theta_3^2 \cos q_2 \sin(q_2)\, \dot{q}_1^2$$
$$-\theta_2 \theta_4 g \cos q_1 + \theta_3 \theta_5 g \cos q_2 \cos(q_1 + q_2)$$

thus

$$\ddot{q}_1 = \frac{1}{\theta_1 \theta_2 - \theta_3^2 \cos^2 q_2} \left[\theta_2 \tau_1 + F(q_1, \dot{q}_1, q_2, \dot{q}_2) \right] \qquad (5.24)$$

Introducing the above in (5.22), one has

$$\dot{V} = \dot{q}_1 \left[\tau_1 \left(k_E \tilde{E} + \frac{k_D \theta_2}{\theta_1 \theta_2 - \theta_3^2 \cos^2 q_2} \right) + \frac{k_D F(q_1, \dot{q}_1, q_2, \dot{q}_2)}{\theta_1 \theta_2 - \theta_3^2 \cos^2 q_2} + k_P \tilde{q}_1 \right]$$

We propose a control law such that

$$\tau_1 \left(k_E \tilde{E} + \frac{k_D \theta_2}{\theta_1 \theta_2 - \theta_3^2 \cos^2 q_2} \right) + \frac{k_D F(q_1, \dot{q}_1, q_2, \dot{q}_2)}{\theta_1 \theta_2 - \theta_3^2 \cos^2 q_2} + k_P \tilde{q}_1 = -\dot{q}_1$$

$$(5.25)$$

which will lead to

$$\dot{V} = -\dot{q}_1^2 \tag{5.26}$$

The control law in (5.25) will have no singularities, provided that

$$\left(k_E \tilde{E} + \frac{k_D \theta_2}{\theta_1 \theta_2 - \theta_3^2 \cos^2 q_2} \right) \neq 0 \tag{5.27}$$

Note from (5.13) that $\tilde{E} \geq -2(\theta_4 + \theta_5)g$. Thus, (5.27) always holds if the following inequality is satisfied

$$\frac{k_D \theta_2}{\max_{q_2}(\det(D(q)))} > 2k_E(\theta_4 + \theta_5)g \tag{5.28}$$

This gives the following lower bound for $\frac{k_D}{k_E}$

$$\frac{k_D}{k_E} > 2\theta_1(\theta_4 + \theta_5)g \tag{5.29}$$

Note that when using the control law (5.25), the pendulum can get stuck at any equilibrium point in (5.16). In order to avoid any singular points other than E_{top}, we require

$$\left| \tilde{E} \right| < \min \left(|E_{top} - E_{mid}|, |E_{top} - E_{l_1}|, |E_{top} - E_{l_2}| \right) \tag{5.30}$$

$$= \min \left(2\theta_4 g, 2\theta_5 g \right) = c \tag{5.31}$$

Since V is a non-increasing function (see (5.26)), (5.31) will hold if the initial conditions are such that

$$V(0) < k_E \frac{c^2}{2} \tag{5.32}$$

The above defines the region of attraction as will be shown in the next section.

Finally, with this condition, the control law can be written as

$$\tau_1 = \frac{-k_D F\left(q_1, \dot{q}_1, q_2, \dot{q}_2\right) - \left(\theta_1 \theta_2 - \theta_3^2 \cos^2 q_2\right)\left(\dot{q}_1 + k_P \tilde{q}_1\right)}{\left(\theta_1 \theta_2 - \theta_3^2 \cos^2 q_2\right) k_E \tilde{E} + k_D \theta_2} \tag{5.33}$$

The main result is stated in the following theorem.

Theorem 5.1 *Consider the pendubot system (5.6) and the Lyapunov function candidate (5.21), with strictly positive constants k_E, k_D and k_P satisfying (5.29). Provided that the state initial conditions (5.31) and (3.28) are satisfied, then the solution of the closed-loop system with the control law (5.33) converges to the invariant set M given by the homoclinic orbit (5.20) with $(q_1, \dot{q}_1) = (\frac{\pi}{2}, 0)$ and the interval $(q_1, \dot{q}_1, q_2, \dot{q}_2) = (\frac{\pi}{2} - \varepsilon, 0, \varepsilon, 0)$, where $|\varepsilon| < \varepsilon^*$ and ε^* is arbitrarily small.* ∎

The proof will be developed in the following section in which the stability will be analyzed.

5.6   Stability analysis

The stability analysis will be based on LaSalle´s invariance theorem (see for instance [46], page 117). In order to apply LaSalle´s theorem, we are required to define a compact (closed and bounded) set Ω with the property that every solution of system (5.6) that starts in Ω remains in Ω for all future time. Since $V(q, \dot{q})$ in (5.21) is a non-increasing function, (see (5.26)), then q_1, $\dot{q}_1$, and $\dot{q}_2$ are bounded. Since $\cos q_2$, $\sin q_2$, $\cos q_1$, $\sin q_1$, $\cos(q_1 + q_2)$, $\sin(q_1 + q_2)$ are bounded functions, we can define the state z of the closed-loop system as being composed of q_1, $\sin q_1$, $\sin(q_1 + q_2)$, $\dot{q}_1$, $\cos q_2$, $\sin q_2$ and $\dot{q}_2$. Therefore, the solution of the closed-loop system $\dot{z} = F(z)$ remains inside a compact set Ω that is defined by the initial state values. Let Γ be the set of all points in Ω such that $\dot{V}(z) = 0$. Let M be the largest invariant set in Γ. LaSalle´s theorem ensures that every solution starting in Ω approaches M as $t \to \infty$. Let us now compute the largest invariant set M in Γ.

In the set Γ (see (5.26)), $\dot{V} = 0$ and $\dot{q}_1 = 0$, which implies that q_1 and V are constant. From (5.21), it follows that E is also constant. Comparing (5.24) and (5.33), it follows that the control law has been chosen such that

$$-\dot{q}_1 = k_E \tilde{E} \tau_1 + k_D \ddot{q}_1 + k_P \tilde{q}_1 \qquad (5.34)$$

From the above equation, we conclude that $\tilde{E}\tau_1$ is constant in Γ. Since E is also constant, then $\tilde{E}$ is constant and we either have a) $\tilde{E} = 0$ or b) $\tilde{E} \neq 0$. On the other hand, if $\tilde{E} = 0$ then from (5.34) $\tilde{q}_1 = 0$, which means that the three conditions c_1, c_2 and c_3 are satisfied and therefore the trajectory belongs to the homoclinic orbit (5.20). If $\tilde{E} \neq 0$ and since $\tilde{E}\tau_1$ is constant, then τ_1 is also constant.

We will give below a mathematical proof of the fact that the solutions converge to the invariant set M when $\dot{q}_1 = 0$, $\ddot{q}_1 = 0$, q_1 is constant, $\tilde{E}$ is constant and τ_1 is constant.

Proof 5.1 *With $\dot{q}_1 = 0$, $\ddot{q}_1 = 0$, q_1 constant, $\tilde{E}$ constant and τ_1 constant, the system (5.6) becomes*

$$
\begin{aligned}
\tau_1 &= (\theta_2 + \theta_3 \cos q_2)\, \ddot{q}_2 - \theta_3 \sin{(q_2)}\, \dot{q}_2^2 + \theta_4 g \cos q_1 \\
&\quad \theta_5 g \cos{(q_1 + q_2)} && (5.35) \\
0 &= \theta_2 \ddot{q}_2 + \theta_5 g \cos{(q_1 + q_2)} && (5.36)
\end{aligned}
$$

Introducing (5.36) into (5.35), we obtain

$$
\frac{-\theta_3 \theta_5 g}{\theta_2} \cos{(q_1 + q_2)} \cos{(q_2)} - \theta_3 \sin{(q_2)}\, \dot{q}_2^2 + \theta_4 g \cos q_1 = \tau_1 \quad (5.37)
$$

Moreover, the energy E in (5.13) is constant and is given by

$$
E = \frac{1}{2}\theta_2 \dot{q}_2^2 + \theta_4 g \sin q_1 + \theta_5 g \sin(q_1 + q_2) = E_0 \quad (5.38)
$$

Combining the above and (5.37) yields

$$
\begin{aligned}
\tau_1 &= \frac{-\theta_3 \theta_5 g}{\theta_2} \cos{(q_1 + q_2)} \cos{(q_2)} - \frac{2\theta_3}{\theta_2} E_0 \sin{(q_2)} \\
&\quad + \frac{2\theta_3}{\theta_2} \sin{(q_2)}\, (\theta_4 g \sin q_1 + \theta_5 g \sin(q_1 + q_2)) \\
&\quad + \theta_4 g \cos q_1 && (5.39)
\end{aligned}
$$

Differentiating (5.39), we obtain

$$
\begin{aligned}
0 &= \frac{3\theta_3 \theta_5 g}{\theta_2} [\sin(q_1 + q_2)\cos(q_2)\dot{q}_2 + \cos(q_1 + q_2)\sin(q_2)\dot{q}_2] \\
&\quad - \frac{2\theta_3 E_0}{\theta_2} \cos(q_2)\dot{q}_2 + \frac{2\theta_3}{\theta_2}\cos(q_2)\dot{q}_2(\theta_4 g \sin q_1) && (5.40)
\end{aligned}
$$

Thus, we have either the case a) $\dot{q}_2 \neq 0$ or the case b) $\dot{q}_2 = 0$. Let us study each case separately.

- *Case a:* If $\dot{q}_2 \neq 0$, (5.40) becomes

$$0 = \frac{3\theta_3\theta_5 g}{\theta_2} \sin(q_1 + 2q_2) - \frac{2\theta_3 E_0}{\theta_2} \cos(q_2)$$
$$+ \frac{2\theta_3}{\theta_2} \cos(q_2)(\theta_4 g \sin q_1) \tag{5.41}$$

Taking the time derivative of the above (5.41), gives us

$$0 = \frac{6\theta_3\theta_5 g}{\theta_2} \cos(q_1 + 2q_2)\dot{q}_2 + \frac{2\theta_3 E_0}{\theta_2} \sin(q_2)\dot{q}_2$$
$$- \frac{2\theta_3}{\theta_2} \sin(q_2)\dot{q}_2(\theta_4 g \sin q_1) \tag{5.42}$$

Since $\dot{q}_2 \neq 0$, we divide the above by $\dot{q}_2$ and differentiating the equation again yields

$$0 = -\frac{12\theta_3\theta_5 g}{\theta_2} \sin(q_1 + 2q_2)\dot{q}_2 + \frac{2\theta_3 E_0}{\theta_2} \cos(q_2)\dot{q}_2$$
$$- \frac{2\theta_3}{\theta_2} \cos(q_2)\dot{q}_2(\theta_4 g \sin q_1) \tag{5.43}$$

Dividing (5.43) by $\dot{q}_2$ yields

$$0 = -\frac{12\theta_3\theta_5 g}{\theta_2} \sin(q_1 + 2q_2) + \frac{2\theta_3 E_0}{\theta_2} \cos(q_2)$$
$$- \frac{2\theta_3}{\theta_2} \cos(q_2)(\theta_4 g \sin q_1) \tag{5.44}$$

Combining (5.41) and (5.44), it follows that

$$\sin(q_1 + 2q_2) = 0 \tag{5.45}$$

Therefore, $q_1 + 2q_2$ is constant. Since q_1 is constant, then q_2 is also constant and $\dot{q}_2 = 0$. This contradicts the premise, i.e. $\dot{q}_2 \neq 0$. This particular case b) is not possible. The only case is finally $\dot{q}_2 = 0$.

- *Case b:* If $\dot{q}_2 = 0$ then $\ddot{q}_2 = 0$ and q_2 is constant. From (5.36), we have $\cos(q_1 + q_2) = 0$ and from (5.35), we then have

$$\theta_4 g \cos q_1 = \tau_1 \tag{5.46}$$

Moreover, from (5.34), we have

$$k_E \tilde{E} \tau_1 + k_P \tilde{q}_1 = 0 \tag{5.47}$$

Since $\cos(q_1 + q_2) = 0$, *then* $\sin(q_1 + q_2) = \pm 1$ *and the energy* $\tilde{E}$ *is given by*

$$\tilde{E} = \begin{cases} \theta_4 g(\sin q_1 - 1) & if \quad \sin(q_1 + q_2) = 1 \\ \theta_4 g(\sin q_1 - 1) - 2\theta_5 g & if \quad \sin(q_1 + q_2) = -1 \end{cases} \quad (5.48)$$

Introducing the above in (5.47) and using (5.46), we obtain if $\sin(q_1 + q_2) = 1$

$$\frac{k_E \theta_4^2 g^2}{k_P}(1 - \sin q_1)\cos q_1 = \tilde{q}_1 \quad (5.49)$$

and if $\sin(q_1 + q_2) = -1$

$$\frac{k_E \theta_4 g}{k_P}(\theta_4 g(1 - \sin q_1) + 2\theta_5 g)\cos q_1 = \tilde{q}_1 \quad (5.50)$$

In both cases, we can always choose some particular constants k_E *and* k_P, *such that the only solution of the above equations (5.49) and (5.50) is* $q_1 = \frac{\pi}{2}$. *Looking at both graphics of the above equations (5.49) and (5.50), it can be easily seen that if* $\frac{k_E}{k_P}$ *is not too big* ($\frac{k_E}{k_P} < 25$), *the only solution is* $q_1 = \frac{\pi}{2}$. *We finally get another constraint on the coefficients.*

Provided that the appropriate constraint on $\frac{k_E}{k_P}$ *is satisfied, we ensure that* $q_1 = \frac{\pi}{2}$. *Therefore, since* $\cos(q_1 + q_2) = 0$, *it follows that* $q_2 = 0[\pi]$. *Furthermore, in view of the constraints imposed on the initial conditions, the position when* $q_2 = \pi$ *is excluded. Therefore* $q_2 = 0$ *and finally* $\tilde{E} = 0$. *This contradicts the assumption* $\tilde{E} \neq 0$ *and thus the only possible case is* $\tilde{E} = 0$. ∎

Finally, the largest invariant set M is given by the homoclinic orbit (5.20) with $(q_1, \dot{q}_1) = (\frac{\pi}{2}, 0)$ and the interval $(q_1, \dot{q}_1, q_2, \dot{q}_2) = (\frac{\pi}{2} - \varepsilon, 0, \varepsilon, 0)$, where $|\varepsilon| < \varepsilon^*$ and ε^* is arbitrarily small. Provided that the state initial conditions satisfy (5.31) and (5.32), and $k_P > 0$ is sufficiently small, then all the solutions converge to the invariant set M. This ends the proof of Theorem 5.1.

5.7 Simulation results

In order to observe the performance of the proposed control law based on passivity, we performed simulations on MATLAB using SIMULINK.

We considered the system taking parameters $\theta_{i,1\leq i\leq 5}$ of a real pendubot from the University of Illinois, i.e. $\theta_1 = 0.034$, $\theta_2 = 0.0125$, $\theta_3 = 0.01$, $\theta_4 = 0.215$ and $\theta_5 = 0.073$. We chose the gains $k_P = 1$, $k_D = 1$ and $k_E = 5$, to increase the convergence rate in order to switch to a linear stabilizing controller in a reasonable time.

Our algorithm allows us to bring the pendubot close to the top position, but the second link will remain swinging while getting closer and closer to the top position. Once the system is close enough to the top position, i.e. $(|x| \leq 0.2, |\dot{x}| \leq 0.2, |\theta| \leq 0.3, |\dot{\theta}| \leq 0.2)$, we have switched to a linear LQR controller $\tau_1 = -K[q_1 \quad \dot{q}_1 \quad q_2 \quad \dot{q}_2]^T$ where $K = [16.46 \quad 3.13 \quad 16.24 \quad 2.07]$.

Figures 5.2 and 5.3 show the results for an initial position

$$q_1 = 0 \qquad\qquad q_2 = 0.4$$
$$\dot{q}_1 = 0.1 \qquad\qquad \dot{q}_2 = 0.1$$

Simulations showed that our non-linear control law brings the state of the system to the homoclinic orbit (see the phase plot in Figure 5.3). The first link q_1 converges to $\frac{\pi}{2}$. $\tilde{E}$ goes to zero, i.e. the energy E goes to the energy at the top position E_{top}. Note that the control law τ_1 is different from zero. Switching to the linear controller occurs approximately at time $t = 160\ s$. Then, the system stops at its unstable top position and the control law is going to zero.

5.8   Experimental results

In addition to real-time experiments on the inverted pendulum, we performed experiments on a pendubot setting at the University of Illinois at Urbana-Champaign. The parameters of the model used for the controller design and the linear controller gains K are the same as in the previous section. For this experiment, we chose the gains $k_P = 10$, $k_D = 1$ and $k_E = 32$.

Figure 5.4 shows the results for an initial position

$$q_1 = -\frac{\pi}{2} + \epsilon \qquad\qquad q_2 = \epsilon$$
$$\dot{q}_1 = \epsilon \qquad\qquad \dot{q}_2 = \epsilon$$

where ϵ is small.

Real-time experiments showed that the non-linear control law brings the system to the homoclinic orbit (see the phase plot in Figure 5.4).

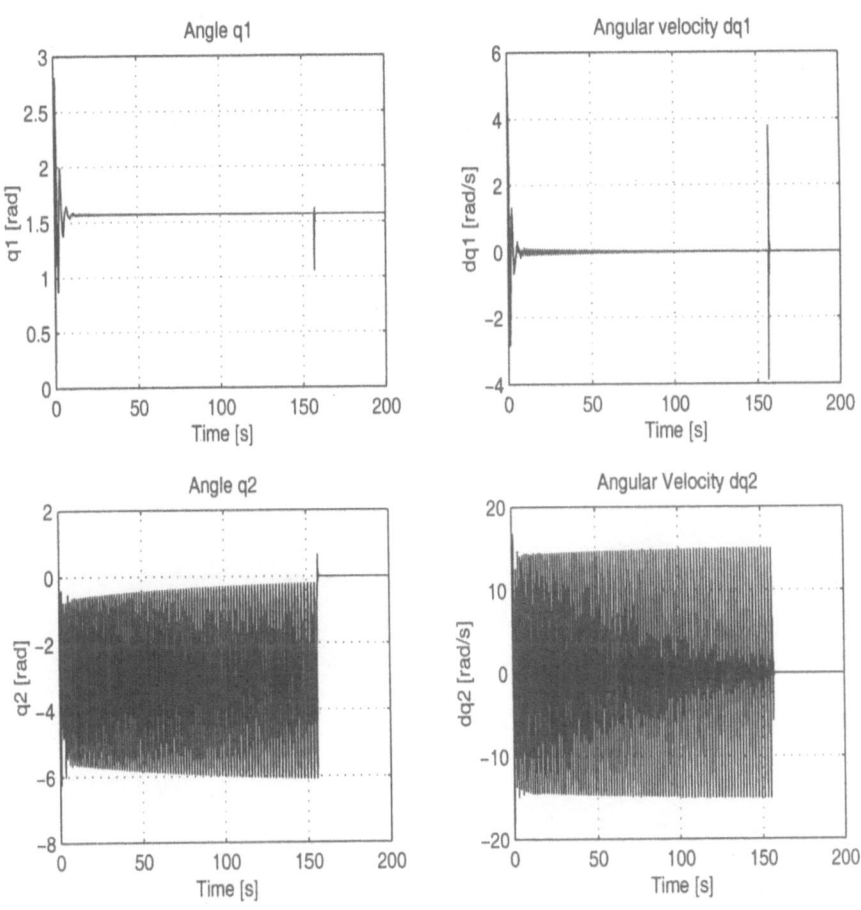

Figure 5.2: Simulation results

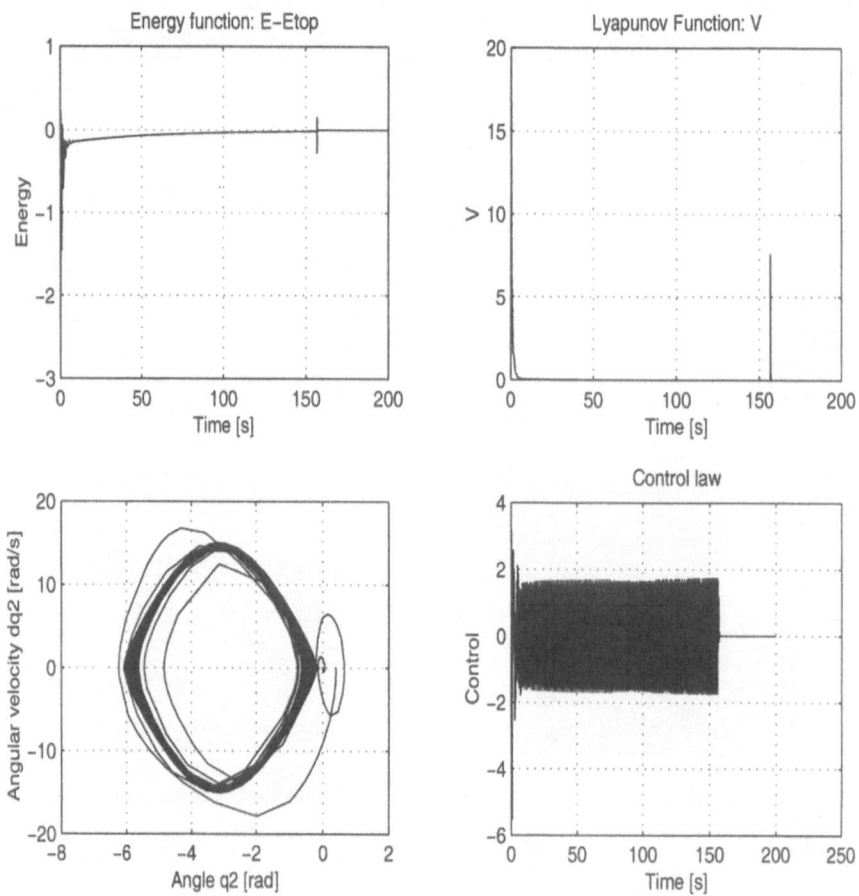

Figure 5.3: Simulation results

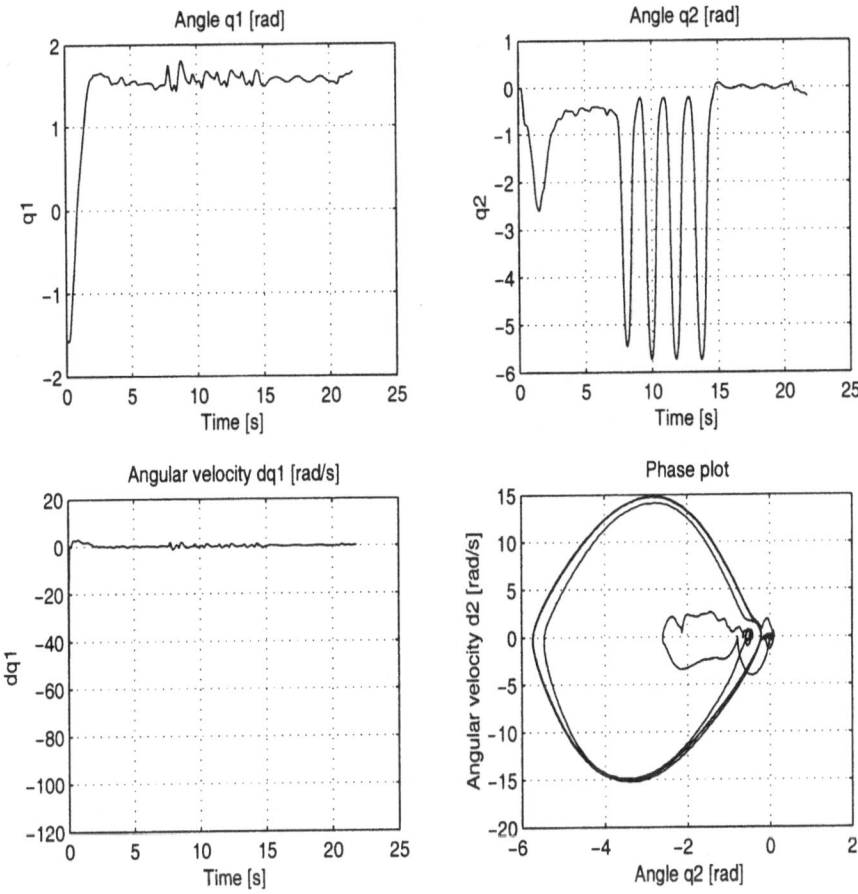

Figure 5.4: Experimental results

Indeed, in Figure 5.4, the plotting of q_2 shows that at time $t = 10$ s, the second link is swinging while getting closer and closer to the top position. Once the system is close enough to this point, switching to the linear controller occurs at time $t = 15$ s. Note that the initial conditions and condition (5.29) lie slightly outside the domain of attraction. This proves that the domain of attraction is conservative.

5.9 Conclusions

We have presented a control strategy for the pendubot that brings the state either arbitrarily close to the top position or to a homoclinic orbit

that will eventually enter the basin of attraction of any locally convergent controller. The control strategy is based on an energy approach
and the passivity properties of the pendubot. A Lyapunov function is
obtained using the total energy of the system. The analysis is carried
out using LaSalle's theorem.

It has been proved that the first link converges to the upright position
while the second oscillates and converges to the homoclinic orbit. This
has also been observed in simulations and tested in a real pendubot.

In order to compare our controller with the one proposed by [107],
we can remark that in our approach the control input amplitude does
not need to be very large since at every cycle (of the second link) we
are only required to slightly increase the energy. In other words, we do
not need high gain controllers.

Chapter 6

The Furuta pendulum

6.1 Introduction

The inverted pendulum is a very popular experiment used for educational purposes in modern control theory and this system can appear with different constructions. As we have seen in Chapter 3, the structure of the conventional inverted pendulum is the rail-cart type, which consists of a cart running on a rail and a pendulum attached to the cart. The inverted pendulum of this type has the movement limitation of its cart as a restriction of the control system. On the other hand, the Furuta pendulum has a different structure. It has a direct-drive motor as its actuator source and its pendulum attached to the rotating shaft of the motor. This inverted pendulum on the rotating arm was first developed by K. Furuta at Tokyo Institute of Technology. The product of the experiment was called the TITech pendulum (see [30, 43, 123]).

Since the angular acceleration of the pole cannot be controlled directly, the Furuta pendulum is an underactuated mechanical system. Therefore, the techniques developed for fully actuated mechanical robot manipulators cannot be used to control the Furuta pendulum.

In 1992, Furuta et al. [30] proposed a robust swing-up control using a subspace projected from the whole state space. Their controller uses a bang-bang pseudo-state feedback control method. In 1995, Yamakita et al. [123] considered different methods to swing up a double pendulum. One is based on an energy approach and another one is based on a robust control method. In 1996, Iwashiro et al. [43] considered a golf shot with a rotational (Furuta) pendulum using control methods based on an energy approach. Olfati-Saber [77] proposed in 1999 semi-global stabilization for the rotational inverted (or Furuta) pendulum using fixed point controllers as for the cart-pole system (see the introduction in Chapter 3). Then, in 2000, he introduced new cascade normal forms for underactuated mechanical systems in [78]. The main benefit of this transformation was to reduce the overall system to control a lower order non-linear subsystem in the normal form. He illustrated his result with the example of the rotational pendulum. Contrary to the technique proposed here, the magnitude of the control input in his scheme increases as the initial state is far from the origin.

The stabilization algorithm proposed here is again an adaptation of the technique presented in the previous chapters and is also developed in [19]. We will consider the passivity properties of the Furuta pendulum and use an energy-based approach to establish the proposed control law. The control algorithm's convergence analysis is based on Lyapunov theory.

In Section 6.2, we present the model of the Furuta pendulum obtained using Euler-Lagrange equations. We also establish its passivity properties. The control law is developed in Section 6.4 and the stability analysis of the closed-loop system is given in Section 6.5. Simulations are presented in Section 6.6 and conclusions are finally given in Section 6.7.

6.2   Modeling of the system

The Furuta pendulum is different to the conventional cart-pole inverted pendulum. The Furuta pendulum requires less space and has fewer unmodelled dynamics owing to a power transmission mechanism, since the shaft around which the pendulum is rotated is directly attached to the motor shaft. The coordinate system and notations are described in Figure 6.1. We will assume that the friction is negligible.

I_0 :   Inertia of the arm
L_0 :   Total length of the arm
m_1 :   Mass of the pendulum
l_1 :   Distance to the center of gravity of the pendulum
J_1 :   Inertia of the pendulum around its center of gravity
θ_0 :   Rotational angle of the arm
θ_1 :   Rotational angle of the pendulum
τ :   Input torque applied on the arm

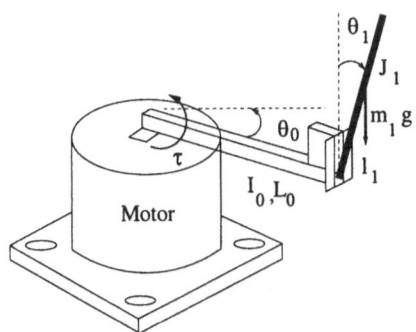

Figure 6.1: The Furuta pendulum system

6.2.1 Energy of the system

The total energy of the system is the sum of the kinetic energy K and the potential energy P of the arm and the pendulum.

The arm

The kinetic energy of the arm is given by

$$K_0 = \frac{1}{2} I_0 \dot{\theta}_0{}^2 \tag{6.1}$$

Its potential energy is null, since no gravitational forces act on the horizontal arm.

The pendulum

The kinetic energy of the pendulum is given by

$$
\begin{aligned}
K_1 = \; & \tfrac{1}{2} J_1 \dot{\theta}_1{}^2 + \tfrac{1}{2} m_1 \Big[\big\{ \tfrac{d}{dt}(L_0 \sin\theta_0 + l_1 \sin\theta_1 \cos\theta_0) \big\}^2 \\
& + \big\{ \tfrac{d}{dt}(L_0 \cos\theta_0 - l_1 \sin\theta_1 \sin\theta_0) \big\}^2 + \big\{ \tfrac{d}{dt}(l_1 \cos\theta_1) \big\}^2 \Big]
\end{aligned}
\tag{6.2}
$$

where the first term corresponds to the kinetic energy due to the angular velocity of the pendulum while the last three terms are due to the tangential velocity, the radial velocity and the vertical velocity of the pendulum respectively. After some simple computations, K_1 reduces to

$$K_1 = \tfrac{1}{2}J_1\dot{\theta_1}^2 + \tfrac{1}{2}m_1 L_0^2 \dot{\theta_0}^2 + \tfrac{1}{2}m_1 l_1^2 \dot{\theta_1}^2$$
$$+ \tfrac{1}{2}m_1 l_1^2 \sin^2\theta_1 \dot{\theta_0}^2 + m_1 L_0 l_1 \cos\theta_1 \dot{\theta_0}\dot{\theta_1} \qquad (6.3)$$

Its potential energy is given by

$$P_1 = m_1 g l_1 (\cos\theta_1 - 1) \qquad (6.4)$$

6.2.2   Euler-Lagrange dynamic equations

The equations of motion can be obtained using an Euler-Lagrange formulation

$$\frac{d}{dt}\left(\frac{\partial L}{\partial \dot{\theta_i}}\right) - \frac{\partial L}{\partial \theta_i} = F_i \qquad (6.5)$$

where $L = K - P$, $K = K_0 + K_1$ and $P = P_1$. We have

$$\left(\frac{\partial L}{\partial \dot{\theta_0}}\right) = \left[I_0 + m_1(L_0^2 + l_1^2\sin^2\theta_1)\right]\dot{\theta_0} + m_1 l_1 L_0 \cos\theta_1 \dot{\theta_1}$$

$$\left(\frac{\partial L}{\partial \theta_0}\right) = 0$$

$$\left(\frac{\partial L}{\partial \dot{\theta_1}}\right) = m_1 l_1 L_0 \cos\theta_1 \dot{\theta_0} + \left[J_1 + m_1 l_1^2\right]\dot{\theta_1}$$

$$\left(\frac{\partial L}{\partial \theta_1}\right) = m_1 l_1^2 \sin\theta_1 \cos\theta_1 \dot{\theta_0}^2 - m_1 l_1 L_0 \sin\theta_1 \dot{\theta_1}\dot{\theta_0} + m_1 g l_1 \sin\theta_1$$

and thus, the system is given by

$$\tau = \left[I_0 + m_1(L_0^2 + l_1^2\sin^2\theta_1)\right]\ddot{\theta_0} + m_1 l_1 L_0 \cos\theta_1 \ddot{\theta_1}$$
$$+ m_1 l_1^2 \sin(2\theta_1)\dot{\theta_0}\dot{\theta_1} - m_1 l_1 L_0 \sin\theta_1 \dot{\theta_1}^2 \qquad (6.6)$$
$$0 = m_1 l_1 L_0 \cos\theta_1 \ddot{\theta_0} + \left[J_1 + m_1 l_1^2\right]\ddot{\theta_1}$$
$$- m_1 l_1^2 \sin\theta_1 \cos\theta_1 \dot{\theta_0}^2 - m_1 g l_1 \sin\theta_1 \qquad (6.7)$$

In compact form, the system can be written as

$$D(q)\ddot{q} + C(q,\dot{q})\dot{q} + g(q) = F \qquad (6.8)$$

where

$$q = \begin{bmatrix} \theta_0 \\ \theta_1 \end{bmatrix} \quad D(q) = \begin{bmatrix} I_0 + m_1(L_0^2 + l_1^2 \sin^2 \theta_1) & m_1 l_1 L_0 \cos \theta_1 \\ m_1 l_1 L_0 \cos \theta_1 & J_1 + m_1 l_1^2 \end{bmatrix} \quad (6.9)$$

$$C(q, \dot{q}) = \begin{bmatrix} \frac{1}{2} m_1 l_1^2 \sin(2\theta_1) \dot{\theta}_1 & -m_1 l_1 L_0 \sin \theta_1 \dot{\theta}_1 + \frac{1}{2} m_1 l_1^2 \sin(2\theta_1) \dot{\theta}_0 \\ -\frac{1}{2} m_1 l_1^2 \sin(2\theta_1) \dot{\theta}_0 & 0 \end{bmatrix}$$
$$(6.10)$$

$$g(q) = \begin{bmatrix} 0 \\ -m_1 g l_1 \sin \theta_1 \end{bmatrix} \quad \text{and} \quad F = \begin{bmatrix} \tau \\ 0 \end{bmatrix} \quad (6.11)$$

Note that $D(q)$ is symmetric and also

$$\begin{aligned} d_{11} &= I_0 + m_1(L_0^2 + l_1^2 \sin^2 \theta_1) & (6.12) \\ &\geq I_0 + m_1 L_0^2 > 0 & (6.13) \end{aligned}$$

and

$$\begin{aligned} \det(D(q)) &= (I_0 + m_1(L_0^2 + l_1^2 \sin^2 \theta_1))(J_1 + m_1 l_1^2) - m_1^2 l_1^2 L_0^2 \cos^2 \theta_1 \\ &= (I_0 + m_1 l_1^2 \sin^2 \theta_1)(J_1 + m_1 l_1^2) + J_1 m_1 L_0^2 \\ &\quad + m_1^2 l_1^2 L_0^2 \sin^2 \theta_1 > 0 \end{aligned} \quad (6.14)$$

Therefore, $D(q)$ is positive definite for all q. From (6.9) and (6.10) it follows that

$$\dot{D}(q) - 2C(q, \dot{q}) = m_1 l_1 (l_1 \sin(2\theta_1) \dot{\theta}_0 - L_0 \sin \theta_1 \dot{\theta}_1) \begin{bmatrix} 0 & -1 \\ 1 & 0 \end{bmatrix} \quad (6.15)$$

which is a skew-symmetric matrix. This constitutes an important property, which will be used in establishing the passivity property of the Furuta pendulum

$$z^T (\dot{D}(q) - 2C(q, \dot{q})) z = 0 \qquad \forall z \qquad (6.16)$$

The potential energy of the system is defined as $P = m_1 g l_1 (\cos \theta_1 - 1)$. Note that P is related to $g(q)$ as follows

$$g(q) = \frac{\partial P}{\partial q} = \begin{bmatrix} 0 \\ -m_1 g l_1 \sin \theta_1 \end{bmatrix} \quad (6.17)$$

6.2.3 Passivity properties of the Furuta pendulum

The total energy of the system is given by

$$E = K(q, \dot{q}) + P(q)$$
$$= \tfrac{1}{2}\dot{q}^T D(q)\dot{q} + m_1 g l_1 (\cos\theta_1 - 1) \tag{6.18}$$

Therefore, from (6.8)-(6.11), (6.15)-(6.17), we obtain

$$\dot{E} = \dot{q}^T D(q)\ddot{q} + \tfrac{1}{2}\dot{q}^T \dot{D}(q)\dot{q} + \dot{q}^T g(q)$$
$$= \dot{q}^T(-C(q,\dot{q})\dot{q} - g(q) + F) + \tfrac{1}{2}\dot{q}^T \dot{D}(q)\dot{q} + \dot{q}^T g(q) \tag{6.19}$$
$$= \dot{q}^T F = \dot{\theta}_0 \tau$$

Integrating both sides of the above equation, we obtain

$$\int_0^t \dot{\theta}_0 \tau dt = E(t) - E(0) \geq -2m_1 g l_1 - E(0) \tag{6.20}$$

Therefore, the system having τ as input and $\dot{\theta}_0$ as output is passive. Note that for $\tau = 0$ and $\theta_0 \in [0, 2\pi[$, the system (6.8) has a subset of two equilibrium set of points; $(\theta_0, \dot{\theta}_0, \theta_1, \dot{\theta}_1) = (*, 0, 0, 0)$ is an unstable equilibrium set of points and $(\theta_0, \dot{\theta}_0, \theta_1, \dot{\theta}_1) = (*, 0, \pi, 0)$ is a stable equilibrium set of points. The total energy $E(q, \dot{q})$ is equal to 0 for the unstable equilibrium set of points and to $-2m_1 g l_1$ for the stable equilibrium set of points. The control objective is to stabilize the system around its unstable equilibrium point $(\theta_0, \dot{\theta}_0, \theta_1, \dot{\theta}_1) = (0, 0, 0, 0)$, i.e. to bring the pendulum to its upper position and the arm angle to zero simultaneously.

6.3 Controllability of the linearized model

When the pendulum is in a neighborhood of its top unstable equilibrium position, a linear controller can stabilize the pendulum quite adequately. In order to implement a balancing linear controller, the general non-linear differential equations (6.6) and (6.7) are linearized about the top equilibrium position. Provided that the linearized system is controllable, we can design a linear controller. Let us therefore compute the rank of the controllability matrix. The general non-linear equations can be rewritten as follows

$$\ddot{\theta}_0 = \frac{1}{\det(D(q))} \Big[(J_1 + m_1 l_1^2)\tau - (J_1 + m_1 l_1^2) m_1 l_1^2 \sin(2\theta_1)\dot{\theta}_0\dot{\theta}_1$$

$$-\frac{1}{2} m_1^2 l_1^3 L_0 \cos\theta_1 \sin(2\theta_1)\dot{\theta}_0^2 + (J_1 + m_1 l_1^2) m_1 l_1 L_0 \sin\theta_1\dot{\theta}_1^2$$

$$- m_1^2 l_1^2 L_0 g \cos\theta_1 \sin\theta_1 \Big] \tag{6.21}$$

$$\ddot{\theta}_1 = \frac{1}{\det(D(q))} \Big[-(m_1 l_1 L_0 \cos\theta_1)\tau - m_1^2 l_1^2 L_0^2 \sin\theta_1 \cos\theta_1\dot{\theta}_1^2$$

$$+ m_1 l_1^2 \sin(2\theta_1)\dot{\theta}_0 \Big[m_1 l_1 L_0 \cos\theta_1\dot{\theta}_1 + \frac{1}{2}(I_0 + m_1 L_0^2 + l_1^2 \sin^2\theta_1)\dot{\theta}_0 \Big]$$

$$+ (I_0 + m_1 L_0^2 + l_1^2 \sin^2\theta_1) m_1 l_1 g \sin\theta_1 \Big] \tag{6.22}$$

Linearizing the non-linear equations about the top unstable equilibrium point, we obtain

$$\frac{d}{dt}\begin{bmatrix} \theta_0 \\ \dot{\theta}_0 \\ \theta_1 \\ \dot{\theta}_1 \end{bmatrix} = \begin{bmatrix} 0 & 1 & 0 & 0 \\ 0 & 0 & \frac{-m_1^2 l_1^2 L_0 g}{I_0(J_1+m_1 l_1^2)+J_1 m_1 L_0^2} & 0 \\ 0 & 0 & 0 & 1 \\ 0 & 0 & \frac{(I_0+m_1 L_0^2)m_1 l_1 g}{I_0(J_1+m_1 l_1^2)+J_1 m_1 L_0^2} & 0 \end{bmatrix}\begin{bmatrix} \theta_0 \\ \dot{\theta}_0 \\ \theta_1 \\ \dot{\theta}_1 \end{bmatrix}$$

$$+ \begin{bmatrix} 0 \\ \frac{J_1+m_1 l_1^2}{I_0(J_1+m_1 l_1^2)+J_1 m_1 L_0^2} \\ 0 \\ \frac{-m_1 l_1 L_0}{I_0(J_1+m_1 l_1^2)+J_1 m_1 L_0^2} \end{bmatrix} \tau = AX + B\tau$$

We then have

$$B = \begin{bmatrix} 0 \\ \frac{J_1+m_1 l_1^2}{I_0(J_1+m_1 l_1^2)+J_1 m_1 L_0^2} \\ 0 \\ \frac{-m_1 l_1 L_0}{I_0(J_1+m_1 l_1^2)+J_1 m_1 L_0^2} \end{bmatrix}, \quad AB = \begin{bmatrix} \frac{J_1+m_1 l_1^2}{I_0(J_1+m_1 l_1^2)+J_1 m_1 L_0^2} \\ 0 \\ \frac{-m_1 l_1 L_0}{I_0(J_1+m_1 l_1^2)+J_1 m_1 L_0^2} \\ 0 \end{bmatrix},$$

$$A^2 B = \begin{bmatrix} 0 \\ \frac{m_1^3 l_1^3 L_0^2 g}{(I_0(J_1+m_1 l_1^2)+J_1 m_1 L_0^2)^2} \\ 0 \\ \frac{-m_1^2 l_1^2 L_0 g(I_0+m_1 L_0^2)}{(I_0(J_1+m_1 l_1^2)+J_1 m_1 L_0^2)^2} \end{bmatrix}, \quad A^3 B = \begin{bmatrix} \frac{m_1^3 l_1^3 L_0^2 g}{(I_0(J_1+m_1 l_1^2)+J_1 m_1 L_0^2)^2} \\ 0 \\ \frac{-m_1^2 l_1^2 L_0 g(I_0+m_1 L_0^2)}{(I_0(J_1+m_1 l_1^2)+J_1 m_1 L_0^2)^2} \\ 0 \end{bmatrix}$$

and det $\left(B|AB|A^2B|A^3B\right) = \frac{m_1^4 l_1^4 L_0^2 g^2}{(I_0(J_1+m_1l_1^2)+J_1m_1L_0^2)^4}$

The linearized system is controllable. Therefore, a full state feedback $f = -K^T X$ with an appropriate gain vector K is able to successfully stabilize the system in a neighborhood of its unstable equilibrium point.

6.4   Stabilization algorithm

Let us first note that in view of (6.18), (6.9) and (6.10), if $\dot\theta_0 = 0$ and $E(q, \dot q) = 0$ then

$$\frac{1}{2}(J_1 + m_1l_1^2)\dot\theta_1^2 = m_1gl_1(1 - \cos\theta_1) \qquad (6.23)$$

The above equation defines a particular trajectory that corresponds to a homoclinic orbit. Note that $\dot\theta_1 = 0$ only when $\theta_1 = 0$. This means that the pendulum angular position moves clockwise or counter-clockwise until it reaches the equilibrium point $(\theta_1, \dot\theta_1) = (0, 0)$. Thus, our objective can be reached if the system can be brought to the orbit (6.23) for $\dot\theta_0 = 0$, $\theta_0 = 0$ and $E = 0$. Bringing the system to this homoclinic orbit solves the problem of "swinging up" the pendulum. In order to balance the pendulum at the upper equilibrium position, the control must eventually be switched to a controller that guarantees (local) asymptotic stability of this equilibrium. By guaranteeing convergence to the above homoclinic orbit, we guarantee that the trajectory will enter the basin of attraction of any (local) balancing controller. We do not consider here the design of the balancing controller in this chapter.

The passivity property of the system suggests us to use the total energy E in (6.18) in the controller design. Since we wish to bring to zero θ_0, $\dot\theta_0$ and E, we propose the following Lyapunov function candidate

$$V(q, \dot q) = \frac{k_E}{2}E(q, \dot q)^2 + \frac{k_\omega}{2}\dot\theta_0^2 + \frac{k_\theta}{2}\theta_0^2 \qquad (6.24)$$

where k_E, k_ω and k_θ are strictly positive constants to be defined later. Note that $V(q, \dot q)$ is a positive semi-definite function. Differentiating V and using (6.19), we obtain

$$\dot{V} = k_E E \dot{E} + k_\omega \dot{\theta}_0 \ddot{\theta}_0 + k_\theta \theta_0 \dot{\theta}_0$$

$$= k_E E \dot{\theta}_0 \tau + k_\omega \dot{\theta}_0 \ddot{\theta}_0 + k_\theta \theta_0 \dot{\theta}_0 \qquad (6.25)$$

$$= \dot{\theta}_0 (k_E E \tau + k_\omega \ddot{\theta}_0 + k_\theta \theta_0)$$

Let us now compute $\ddot{\theta}_0$ from (6.8). The inverse of $D(q)$ can be obtained from (6.9) and (6.14) and is given by

$$D^{-1}(q) = \frac{1}{[\det(D(q))]} \begin{bmatrix} J_1 + m_1 l_1^2 & -m_1 l_1 L_0 \cos \theta_1 \\ -m_1 l_1 L_0 \cos \theta_1 & I_0 + m_1 (L_0^2 + l_1^2 \sin^2 \theta_1) \end{bmatrix} \qquad (6.26)$$

with

$$\det(D(q)) = (I_0 + m_1 l_1^2 \sin^2 \theta_1)(J_1 + m_1 l_1^2) + J_1 m_1 L_0^2 + m_1^2 l_1^2 L_0^2 \sin^2 \theta_1$$

Therefore, from (6.8)-(6.11), we have

$$\begin{bmatrix} \ddot{\theta}_0 \\ \ddot{\theta}_1 \end{bmatrix} = \frac{1}{[\det(D(q))]} \begin{pmatrix} (J_1 + m_1 l_1^2)\tau \\ -(m_1 l_1 L_0 \cos \theta_1)\tau \end{pmatrix}$$

$$-D^{-1}(q) \left(C(q, \dot{q}) \begin{bmatrix} \dot{\theta}_0 \\ \dot{\theta}_1 \end{bmatrix} + g(q) \right) \qquad (6.27)$$

$\ddot{\theta}_0$ can thus be written as

$$\ddot{\theta}_0 = \frac{1}{\det(D(q))} \Big[(J_1 + m_1 l_1^2)\tau - (J_1 + m_1 l_1^2)m_1 l_1^2 \sin(2\theta_1)\dot{\theta}_0 \dot{\theta}_1$$

$$-\frac{1}{2} m_1^2 l_1^3 L_0 \cos \theta_1 \sin(2\theta_1)\dot{\theta}_0^2 + (J_1 + m_1 l_1^2)m_1 l_1 L_0 \sin \theta_1 \dot{\theta}_1^2$$

$$- m_1^2 l_1^2 L_0 g \cos \theta_1 \sin \theta_1 \Big] \qquad (6.28)$$

Defining

$$F(q, \dot{q}) = \Big[-(J_1 + m_1 l_1^2)m_1 l_1^2 \sin(2\theta_1)\dot{\theta}_0 \dot{\theta}_1 - \frac{1}{2} m_1^2 l_1^3 L_0 \cos \theta_1 \sin(2\theta_1)\dot{\theta}_0^2$$

$$+(J_1 + m_1 l_1^2)m_1 l_1 L_0 \sin \theta_1 \dot{\theta}_1^2 - m_1^2 l_1^2 L_0 g \cos \theta_1 \sin \theta_1 \Big] \qquad (6.29)$$

we get

$$\ddot{\theta}_0 = \frac{1}{\det(D(q))} \left[(J_1 + m_1 l_1^2)\tau + F(q, \dot{q}) \right] \qquad (6.30)$$

Introducing the above in (6.25), one has

$$\dot{V} = \dot{\theta}_0 \left[\tau \left(k_E E + \frac{k_\omega (J_1 + m_1 l_1^2)}{\det(D(q))} \right) + \frac{k_\omega F(q, \dot{q})}{\det(D(q))} + k_\theta \theta_0 \right] \qquad (6.31)$$

We propose a control law such that

$$\tau \left(k_E E + \frac{k_\omega (J_1 + m_1 l_1^2)}{\det(D(q))} \right) + \frac{k_\omega F(q, \dot{q})}{\det(D(q))} + k_\theta \theta_0 = -k_\delta \dot{\theta}_0 \qquad (6.32)$$

which will lead to

$$\dot{V} = -k_\delta \dot{\theta}_0^2 \qquad (6.33)$$

Note that other functions $f(\dot{\theta}_0)$ such that $\dot{\theta}_0 f(\dot{\theta}_0) > 0$ are also possible, in the right hand side of (6.32).

The control law in (6.32) will have no singularities, provided that

$$\left(k_E E + \frac{k_\omega (J_1 + m_1 l_1^2)}{\det(D(q))} \right) \neq 0 \qquad (6.34)$$

Note from (6.18) that $E \geq -2m_1 g l_1$. Thus, (6.34) always holds if the following inequality is satisfied

$$\frac{k_\omega (J_1 + m_1 l_1^2)}{\max_{\theta_1} (\det(D(q)))} > k_E (2m_1 g l_1) \qquad (6.35)$$

This gives the following lower bound for $\frac{k_\omega}{k_E}$

$$\frac{k_\omega}{k_E} > 2m_1 g l_1 (I_0 + m_1 l_1^2 + m_1 L_0^2) \qquad (6.36)$$

Note that when using the control law (6.32), the pendulum can still get stuck at the (lower) stable equilibrium point, $(\theta_0, \dot{\theta}_0, \theta_1, \dot{\theta}_1) = (0, 0, \pi, 0)$

for which $\tau = 0$. In order to avoid this singular point, which occurs when $E = -2m_1gl_1$ (see (6.18)), it suffices that the following holds

$$|E| < 2m_1gl_1 \tag{6.37}$$

Since V is a non-increasing function (see (6.33)), (6.37) will hold if the initial conditions are such that

$$V(0) < 2k_E m_1^2 g^2 l_1^2 \tag{6.38}$$

The above defines the region of attraction (see Chapter 3 or [59] for more details).

Finally, the control law can be written as

$$\tau = \frac{-k_\omega F(q, \dot{q}) - \det(D(q)) \left(k_\delta \dot{\theta}_0 + k_\theta \theta_0 \right)}{\det(D(q)) k_E E + k_\omega (J_1 + m_1 l_1^2)} \tag{6.39}$$

with k_E and k_ω satisfying (6.36).

6.5 Stability analysis

The stability analysis will be based on LaSalle´s invariance theorem (see for instance [46], page 117). In order to apply LaSalle´s theorem, we are required to define a compact (closed and bounded) set Ω with the property that every solution of system (6.8) that starts in Ω remains in Ω for all future time. Since $V(q, \dot{q})$ in (6.24) is a non-increasing function, (see (6.33)), then θ_0, $\dot{\theta}_0$, and $\dot{\theta}_1$ are bounded. Since $\cos\theta_0$, $\sin\theta_0$, $\cos\theta_1$ and $\sin\theta_1$ are bounded functions, we can define a state z of the closed-loop system composed of θ_0, $\dot{\theta}_0$, $\cos\theta_1$, $\sin\theta_1$ and $\dot{\theta}_1$. Therefore, the solution of the closed-loop system $\dot{z} = F(z)$ remains inside a compact set Ω that is defined by the initial state values. Let Γ be the set of all points in Ω such that $\dot{V}(z) = 0$. Let M be the largest invariant set in Γ. LaSalle´s theorem ensures that every solution starting in Ω approaches M as $t \to \infty$. Let us now compute the largest invariant set M in Γ.

In the set Γ (see (6.33)), $\dot{V} = 0$ and $\dot{\theta}_0 = 0$, which implies that θ_0 and V are constant. From (6.24), it follows that E is also constant. From (6.25) and (6.31)-(6.32), it follows that the control law has been chosen such that

$$-k_\delta \dot{\theta}_0 = k_E E\tau + k_\omega \ddot{\theta}_0 + k_\theta \theta_0 \qquad (6.40)$$

From the above equation, we conclude that $E\tau$ is constant in Γ. Since E is also constant, we either have a) $E = 0$ or b) $E \neq 0$.

- Case a: If $E = 0$ then, from (6.40), $\theta_0 = 0$. Note that τ in (6.39) is bounded in view of (6.34)-(6.38). Recall that $E = 0$ means that the trajectories are in the homoclinic orbit (6.23). In this case, we conclude that θ_0, $\dot{\theta}_0$ and E converge to zero. Note from (6.29) and (6.39) that τ does not necessarily converge to zero.

- Case b: If $E \neq 0$ and since $E\tau$ is constant, then τ is also constant. However, a force input τ constant and different from zero would lead to a contradiction. We will give below a mathematical proof of the fact that if $E \neq 0$ then $\tau = 0$ in Γ.

Proof 6.1 *We will prove that when $\theta_0 = 0$, E is constant and $E \neq 0$, and τ is constant, then τ should be zero. From (6.6) and (6.7) we get*

$$m_1 l_1 L_0 \cos\theta_1 \ddot{\theta}_1 - m_1 l_1 L_0 \sin\theta_1 \dot{\theta_1}^2 = \tau \qquad (6.41)$$
$$[J_1 + m_1 l_1^2]\ddot{\theta}_1 - m_1 g l_1 \sin\theta_1 = 0 \qquad (6.42)$$

Moreover, the energy E (6.18) is constant and given by

$$E = \frac{1}{2}(J_1 + m_1 l_1^2)\dot{\theta}_1^2 + m_1 g l_1 (\cos\theta_1 - 1) \overset{\triangle}{=} E_1 \qquad (6.43)$$

Introducing (6.42) in (6.41), we obtain

$$\sin\theta_1(a\cos\theta_1 - \dot{\theta}_1^2) = \frac{\tau}{b} \qquad (6.44)$$

with $a = \frac{m_1 g l_1}{J_1 + m_1 l_1^2}$ and $b = m_1 l_1 L_0$. The expression (6.43) gives us

$$\dot{\theta}_1^2 = E_2 + c(1 - \cos\theta_1) \qquad (6.45)$$

with $E_2 = \frac{2E_1}{J_1 + m_1 l_1^2}$ and $c = \frac{2m_1 g l_1}{J_1 + m_1 l_1^2}$. Combining (6.45) and (6.44) yields

$$\sin\theta_1((a+c)\cos\theta_1 + d) = \frac{\tau}{b} \qquad (6.46)$$

with $d = -(E_2 + c)$. *Taking the time derivative of (6.46), we obtain*

$$\dot{\theta}_1\left((a+c)\left(\cos^2\theta_1 - \sin^2\theta_1\right) + d\cos\theta_1\right) = 0 \qquad (6.47)$$

If $\dot{\theta}_1 = 0$, then $\ddot{\theta}_1 = 0$ and from (6.42) we conclude that $\sin\theta_1 = 0$. If $\dot{\theta}_1 \neq 0$, then (6.47) becomes

$$(a+c)\left(\cos^2\theta_1 - \sin^2\theta_1\right) + d\cos\theta_1 = 0 \qquad (6.48)$$

Differentiating (6.48), it follows that

$$-\dot{\theta}_1\sin\theta_1\left[4(a+c)\cos\theta_1 + d\right] = 0$$

If $\cos\theta_1 = \frac{-d}{4(a+c)}$ then θ_1 is constant, which implies $\dot{\theta}_1 = 0$, and so $\sin\theta_1 = 0$ (see (6.42)).

In each possible case, we conclude that $\sin\theta_1 = 0$. Then, $\dot{\theta}_1 = 0$. From (6.41) it follows that $\tau = 0$. ∎

We therefore conclude that $\tau = 0$ in Γ. From (6.40) it then follows that $\theta_0 = 0$ in Γ. It only remains to be proved that $E = 0$ when $\theta_0 = 0$, $\dot{\theta}_0 = 0$ and $\tau = 0$. Since $\sin\theta_1 = 0$, it follows that $\theta_0 = 0 \pmod{\pi}$, since $\theta_0 = \pi \pmod{2\pi}$ has been excluded by imposing condition (6.38). Therefore, $\theta_0 = 0 \pmod{2\pi}$, $\dot{\theta}_0 = 0$, $\dot{\theta}_1 = 0$ imply that $E = 0$. This contradicts the assumption $E \neq 0$ in case b) and thus the only possible case is $E = 0$.

The main result can be summarized in the following theorem.

Theorem 6.1 *Consider the Furuta pendulum system (6.6)-(6.7) and the controller in (6.39) with strictly positive constants k_E, k_ω, k_θ and k_δ satisfying (6.36). Provided that the state initial conditions satisfy inequality (6.38), then the solution of the closed-loop system converges to the invariant set M given by the homoclinic orbit (6.23) with $(\theta_0, \dot{\theta}_0) = (0, 0)$. Note that τ does not necessarily converge to zero.* ∎

Remark 6.1 *The above result is local in the sense that the system initial state should belong to the domain of attraction defined in (6.38). However, the same result will hold for arbitrary initial conditions except for a particular manifold bringing the system to the stable equilibrium position with $\theta_1 = \pi$.* ∎

6.6   Simulation results

Figures 6.2 and 6.3 show the performance of the proposed control law. The initial position is

$$\theta_0 = -\frac{\pi}{2} \qquad \dot{\theta}_0 = 0$$

$$\theta_1 = \frac{2.5\pi}{3} \qquad \dot{\theta}_1 = 0$$

and the parameters are $I_0 = 1.75 \times 10^{-2}$, $L_0 = 0.215$, $m_1 = 5.38 \times 10^{-2}$, $l_1 = 0.113$ and $J_1 = 1.98 \times 10^{-4}$. The gains were chosen as $k_E = 480$, $k_\theta = 1$, $k_\omega = 1$ and $k_\delta = 1$.

6.7   Conclusions

We have proposed a control strategy to "swing up" the Furuta pendulum. The control design is based on the passivity properties of this rotational inverted pendulum. Convergence of the trajectories of the system to a homoclinic orbit has been proved by using LaSalle's invariance theorem. We have presented simulations showing the performance of the proposed strategy.

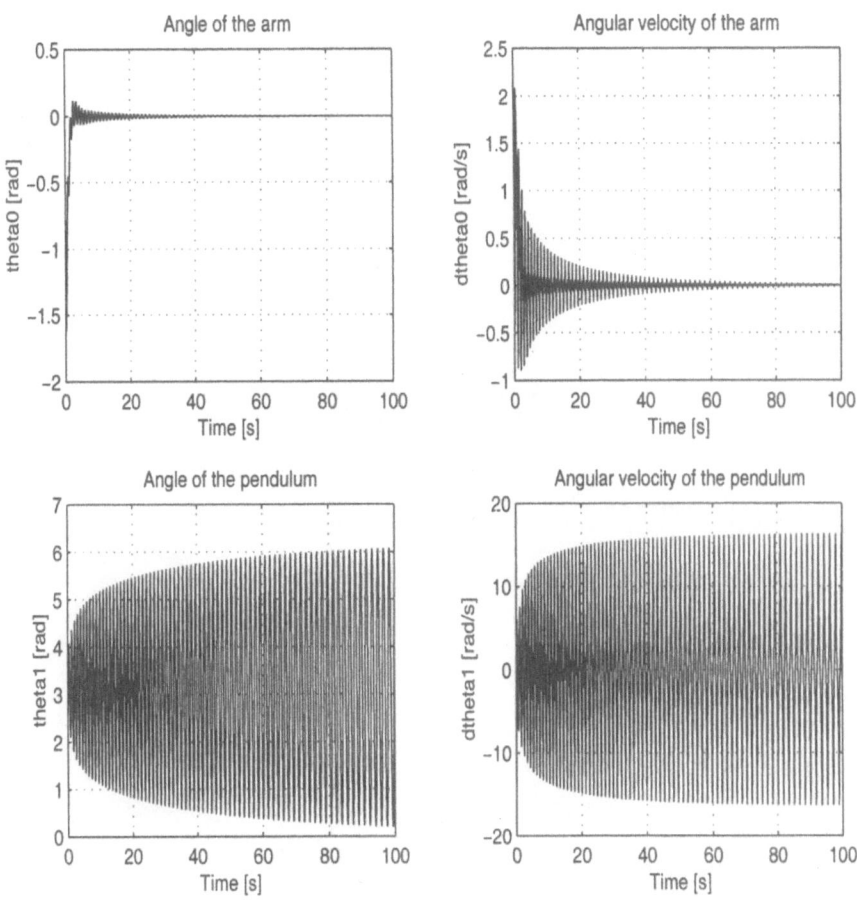

Figure 6.2: Simulation results

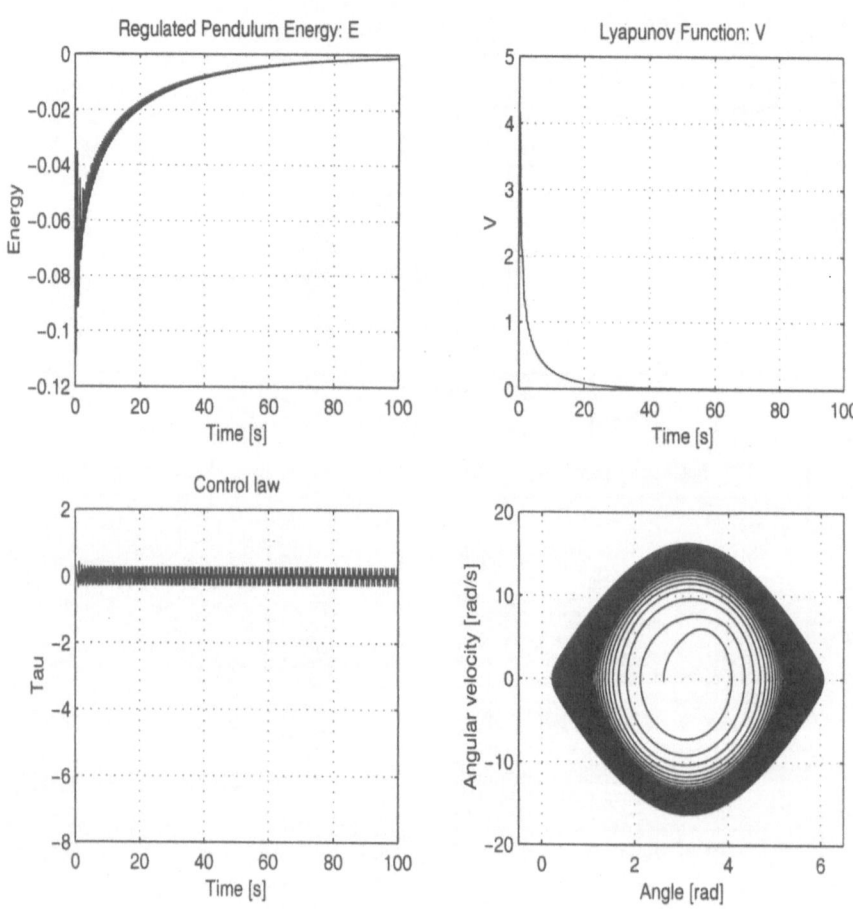

Figure 6.3: Simulation results

Chapter 7

The reaction wheel pendulum

7.1   Introduction

The reaction wheel pendulum is one of the simplest non-linear under-actuated systems. It is a pendulum with a rotating wheel at the end, which is free to spin about an axis parallel to the axis of rotation of the pendulum (see Figure 7.1). The wheel is actuated by a DC-motor, while the pendulum is unactuated. The coupling torque generated by the angular acceleration of the disk can be used to actively control the system. This mechanical system was introduced and studied in [108], where a partial feedback linearization control law was presented.

In [76], Olfati-Saber transformed the reaction wheel (or inertia wheel) pendulum's dynamics into a cascade non-linear system in strict feedback form, using a global change of coordinates in an explicit form. Then, he proposed global asymptotic stabilization of the upright equilibrium point using the standard backstepping procedure. In his approach, contrary to the strategy proposed here, the magnitude of the control input increases with the norm of the state initial condition.

The control objective here will also be to swing the pendulum up and balance it about its unstable inverted position. We will focus our study on the swinging-up control law. The non-linear swinging-up controller will be based on the total energy of the system. The control design will exploit the passivity property of the complete Lagrangian system dynamics. Note that the technique has been presented in [23]. Similar control strategies have been used to control other underactuated mechanical systems in [59] for the cart-pole system, in [24] for the pendubot and in [21] for planar manipulators with springs.

In this chapter, we present two approaches based on the total energy stored in the system. We make use of LaSalle's theorem to prove that the system trajectories asymptotically converge to a homoclinic orbit in both approaches. Therefore, asymptotically, after every swing of the pendulum, the system state gets successively closer to the origin.

The first approach proposed here is such that the wheel's angular velocity converges to zero but does not necessarily bring the wheel to a desired angular position. Nevertheless, the control input can be made smaller than any arbitrary upper bound. The second approach is such that the wheel's angular position converges to zero.

In Section 7.2, we develop the equations of motion of the reaction wheel pendulum. In Sections 7.3 and 7.4, two different energy-based control algorithms are presented. Simulation results are given in Section 7.5. The concluding remarks are presented in Section 7.6.

7.2   The reaction wheel pendulum

7.2.1   Equations of motion

The reaction wheel pendulum is a two-degree-of-freedom robot as shown in Figure 7.1. The pendulum constitutes the first link, while the rotating wheel is the second one. The angle of the pendulum is q_1 and is measured clockwise from the vertical. The angle of the wheel is q_2.

The parameters of the system are described in the following table.

m_1	:	Mass of the pendulum
m_2	:	Mass of the wheel
l_1	:	Length of the pendulum
l_{c1}	:	Distance to the center of mass of the pendulum
I_1	:	Moment of inertia of the pendulum
I_2	:	Moment of inertia of the wheel
q_1	:	Angle that the pendulum makes with the vertical
q_2	:	Angle of the wheel
τ	:	Motor torque input applied on the disk

We introduce the parameter $\bar{m} = m_1 l_{c1} + m_2 l_1$, which will be used later.

The kinetic energy of the pendulum is $K_1 = \frac{1}{2} \left(m_1 l_{c1}^2 + I_1 \right) \dot{q}_1^{\,2}$ and the kinetic energy of the wheel is $K_2 = \frac{1}{2} m_2 l_1^2 \dot{q}_1^{\,2} + \frac{1}{2} I_2 (\dot{q}_1 + \dot{q}_2)^2$. There-

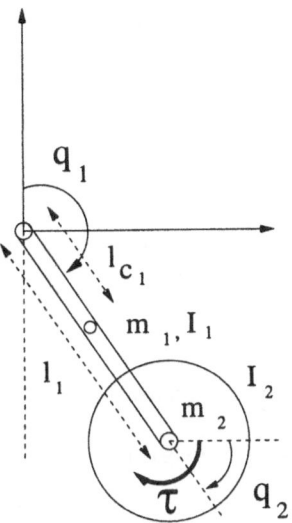

Figure 7.1: The reaction wheel pendulum

fore the total kinetic energy is given by

$$K = K_1 + K_2 = \frac{1}{2}(m_1 l_{c1}^2 + m_2 l_1^2 + I_1 + I_2)\dot{q_1}^2 + I_2 \dot{q_1}\dot{q_2} + \frac{1}{2}I_2\dot{q_2}^2$$

(7.1)

The potential energy of the system is $P = \bar{m}g(\cos(q_1) - 1)$. Finally, the Lagrangian function is given by

$$
\begin{aligned}
L &= K - P \\
L &= \frac{1}{2}(m_1 l_{c1}^2 + m_2 l_1^2 + I_1 + I_2)\dot{q_1}^2 + I_2 \dot{q_1}\dot{q_2} + \frac{1}{2}I_2\dot{q_2}^2 \\
&\quad - \bar{m}g(\cos(q_1) - 1)
\end{aligned}
$$

(7.2)

Using Euler-Lagrange's equations

$$\frac{d}{dt}\left(\frac{\partial L}{\partial \dot{q}}(q, \dot{q})\right) - \frac{\partial L}{\partial q}(q, \dot{q}) = \tau$$

(7.3)

we therefore have

$$\left(\frac{\partial L}{\partial \dot{q}_1}\right) = (m_1 l_{c1}^2 + m_2 l_1^2 + I_1 + I_2)\dot{q}_1 + I_2 \dot{q}_2$$

$$\left(\frac{\partial L}{\partial q_1}\right) = \bar{m}g \sin(q_1)$$

$$\left(\frac{\partial L}{\partial \dot{q}_2}\right) = I_2 \dot{q}_1 + I_2 \dot{q}_2$$

$$\left(\frac{\partial L}{\partial q_2}\right) = 0$$

The dynamic equations of the system are finally given by

$$(m_1 l_{c1}^2 + m_2 l_1^2 + I_1 + I_2)\ddot{q}_1 + I_2 \ddot{q}_2 - \bar{m}g \sin(q_1) = 0 \qquad (7.4)$$

$$I_2 \ddot{q}_1 + I_2 \ddot{q}_2 = \tau \qquad (7.5)$$

In compact form, the system can be rewritten as follows

$$D(q)\ddot{q} + g(q) = u \qquad (7.6)$$

where $q = \begin{bmatrix} q_1 \\ q_2 \end{bmatrix}$ is the vector of generalized coordinates, $u = \begin{bmatrix} 0 \\ \tau \end{bmatrix}$ is the vector of joint torques, $D(q)$ is the inertia matrix and is given by

$$D(q) = \begin{bmatrix} m_1 l_{c1}^2 + m_2 l_1^2 + I_1 + I_2 & I_2 \\ I_2 & I_2 \end{bmatrix} = \begin{bmatrix} d_{11} & d_{12} \\ d_{21} & d_{22} \end{bmatrix} \qquad (7.7)$$

and

$$g(q) = \begin{bmatrix} -\bar{m}g \sin(q_1) \\ 0 \end{bmatrix} \qquad (7.8)$$

Note that the matrix $D(q)$ is constant and positive definite. The equations of motion are also given by

$$d_{11}\ddot{q}_1 + d_{12}\ddot{q}_2 + g(q_1) = 0 \qquad (7.9)$$

$$d_{21}\ddot{q}_1 + d_{22}\ddot{q}_2 = \tau \qquad (7.10)$$

7.2.2 Passivity properties of the system

The total energy of the reaction wheel pendulum is

$$E = \frac{1}{2}\dot{q}^T D(q)\dot{q} + P(q)$$

$$= \frac{1}{2}\dot{q}^T D(q)\dot{q} + \bar{m}g(\cos(q_1) - 1) \qquad (7.11)$$

From (7.6)-(7.8) and (7.11), it follows that

$$\dot{E} = \dot{q}^T D(q)\ddot{q} - \bar{m}g\sin(q_1)\dot{q}_1 = \dot{q}_2\tau \tag{7.12}$$

As a consequence, the system with τ as input and $\dot{q}_2$ as output is passive. The reaction wheel pendulum with zero control input ($\tau = 0$) has an unstable equilibrium at $(q_1, \dot{q}_1, \dot{q}_2) = (0,0,0)$ with energy $E(0,0,0) = 0$ and a stable equilibrium at $(q_1, \dot{q}_1, \dot{q}_2) = (\pi, 0, 0)$ with energy $E(0,0,0) = -2\bar{m}g$. The disk position q_2 can be arbitrary, since the energy in (7.11) does not depend on q_2, i.e. q_2 is a cyclic variable. Hence, the equilibrium points described above are not isolated equilibrium points in the four-dimensional state space of the system.

The control objective will be to control the pendulum position q_1, the pendulum velocity $\dot{q}_1$ and the disk velocity $\dot{q}_2$ and to leave the disk position q_2 unspecified.

Let us consider the state vector $z = [\cos q_1, \sin q_1, \dot{q}_1, \dot{q}_2]^T$. We will bring the state vector z to $[1, 0, 0, 0]^T$.

7.2.3 Linearization of the system

In this section, we will study the controllability of the linearized system about the origin. The equations of motion are given by (see (7.9) and (7.10))

$$\ddot{q}_1 = \frac{d_{22}}{\det(D)}\bar{m}g\sin(q_1) - \frac{d_{12}}{\det(D)}\tau \tag{7.13}$$

$$\ddot{q}_2 = \frac{-d_{21}}{\det(D)}\bar{m}g\sin(q_1) + \frac{d_{11}}{\det(D)}\tau \tag{7.14}$$

where $\det(D) = d_{11}d_{22} - d_{21}d_{12} = (m_1 l_{c1}^2 + m_2 l_1^2 + I_1)I_2$.

Let us consider the vector state $X = [q_1, \dot{q}_1, q_2, \dot{q}_2]$. Differentiating equations (7.13) and (7.14) with respect to the states and evaluating them at the origin leads to the following linear system

$$\frac{d}{dt}\begin{bmatrix} q_1 \\ \dot{q}_1 \\ q_2 \\ \dot{q}_2 \end{bmatrix} = \begin{bmatrix} 0 & 1 & 0 & 0 \\ \frac{d_{22}}{\det(D)}\bar{m}g & 0 & 0 & 0 \\ 0 & 0 & 0 & 1 \\ \frac{-d_{21}}{\det(D)}\bar{m}g & 0 & 0 & 0 \end{bmatrix}\begin{bmatrix} q_1 \\ \dot{q}_1 \\ q_2 \\ \dot{q}_2 \end{bmatrix} + \begin{bmatrix} 0 \\ \frac{-d_{12}}{\det(D)} \\ 0 \\ \frac{d_{11}}{\det(D)} \end{bmatrix}\tau$$

$$= AX + B\tau$$

We then have

$$B = \begin{bmatrix} 0 \\ \frac{-d_{12}}{\det(D)} \\ 0 \\ \frac{d_{11}}{\det(D)} \end{bmatrix} \qquad AB = \begin{bmatrix} \frac{-d_{12}}{\det(D)} \\ 0 \\ \frac{d_{11}}{\det(D)} \\ 0 \end{bmatrix}$$

$$A^2 B = \begin{bmatrix} 0 \\ \frac{-d_{22}d_{12}}{(\det(D))^2}\bar{m}g \\ 0 \\ \frac{d_{21}d_{12}}{(\det(D))^2}\bar{m}g \end{bmatrix} \qquad A^3 B = \begin{bmatrix} \frac{-d_{22}d_{12}}{(\det(D))^2}\bar{m}g \\ 0 \\ \frac{d_{21}d_{12}}{(\det(D))^2}\bar{m}g \\ 0 \end{bmatrix}$$

and $\det\left(B|AB|A^2B|A^3B\right) = \frac{d_{12}^2 m^2 g^2}{(\det(D))^4} = \frac{\bar{m}^2 g^2}{(m_1 l_{c_1}^2 + m_2 l_1^2 + I_1)^4 I_2^2}$. The linearized system is controllable. Therefore, a full state feedback controller $\tau = -K^T X$ with an appropriate gain vector K is able to successfully stabilize the system in a neighborhood of the origin.

7.2.4   Feedback linearization

For completeness purposes, we briefly develop in this section the feedback linearization first presented in [108].

Consider the following output function

$$y = d_{11}q_1 + d_{12}q_2 \tag{7.15}$$

Differentiating (7.15), we obtain

$$\dot{y} = d_{11}\dot{q}_1 + d_{12}\dot{q}_2 \tag{7.16}$$
$$\ddot{y} = \bar{m}g\sin q_1 \tag{7.17}$$
$$y^{(3)} = \bar{m}g\cos(q_1)\dot{q}_1 \tag{7.18}$$
$$y^{(4)} = \bar{m}g\cos(q_1)\ddot{q}_1 - \bar{m}g\sin(q_1)\dot{q}_1^2 \tag{7.19}$$

From (7.9) and (7.10), we get

$$\ddot{q}_1 = \frac{d_{22}}{\det(D)}\bar{m}g\sin q_1 - \frac{d_{12}}{\det(D)}\tau \tag{7.20}$$

Introducing (7.20) in (7.19) yields

$$\begin{aligned} y^{(4)} = {} & \bar{m}g\cos(q_1)\left(-\frac{d_{12}}{\det(D)}\right)\tau \\ & + \bar{m}g\sin(q_1)\left(-\dot{q}_1^2 + \bar{m}g\cos(q_1)\left(\frac{d_{22}}{\det(D)}\right)\right) \end{aligned} \tag{7.21}$$

Thus, the system has a relative degree of four with respect to the output $d_{11}\dot{q}_1 + d_{12}\dot{q}_2$ in the region $-\frac{\pi}{2} < q_1 < \frac{\pi}{2}$, i.e. when $\cos(q_1) \neq 0$. We can define a controller τ, so the closed-loop system is given by

$$y^{(4)} = -\alpha_3 y^{(3)} - \alpha_2 \ddot{y} - \alpha_1 \dot{y} - \alpha_0 y \qquad (7.22)$$

where $s^4 + \alpha_3 s^3 + \alpha_2 s^2 + \alpha_1 s + \alpha_0$ is a stable polynomial. Therefore, $y^{(i)} \to 0$. Finally from (7.16) and (7.17), it follows that $(q_1, \dot{q}_2) \to (0, 0)$. The above shows that the reaction wheel pendulum is feedback linearizable in the region $|q_1| < \frac{\pi}{2}$, i.e. when the pendulum angle q_1 is above the horizontal.

7.3 First energy-based control design

Define the following Lyapunov function candidate

$$V_1 = \frac{1}{2} k_E E^2 + \frac{1}{2} k_v \left(d_{21}\dot{q}_1 + d_{22}\dot{q}_2 \right)^2 \qquad (7.23)$$

where k_E and k_v are strictly positive constants. The time derivative of V_1 is given by

$$
\begin{aligned}
\dot{V}_1 &= k_E E \dot{E} + k_v \left(d_{21}\dot{q}_1 + d_{22}\dot{q}_2 \right) \tau \\
&= \left(k_E E \dot{q}_2 + k_v (d_{21}\dot{q}_1 + d_{22}\dot{q}_2) \right) \tau
\end{aligned}
$$

We propose a controller such that

$$\tau = -k_d \left(k_E E \dot{q}_2 + k_v (d_{21}\dot{q}_1 + d_{22}\dot{q}_2) \right) \qquad (7.24)$$

which leads to

$$\dot{V}_1 = -\frac{1}{k_d} \tau^2 \qquad (7.25)$$

Equations (7.23) and (7.25) imply that V_1 is a non-increasing function, V_1 converges to a constant and $V_1 \leq V_1(0)$. This implies that the energy E remains bounded as well as $\dot{q}_1$ and $\dot{q}_2$. Therefore the closed-loop state vector $z = [\cos q_1, \sin q_1, \dot{q}_1, \dot{q}_2]^T$ is bounded and we can thus apply LaSalle's invariance principle.

In order to apply LaSalle's theorem, we are required to define a compact (closed and bounded) set Ω with the property that every solution of the system $\dot{z} = F(z)$ that starts in Ω remains in Ω for all future time. Therefore, the solutions of the closed-loop system $\dot{z} = F(z)$ remain inside a compact set Ω that is defined by the initial value of z.

Let Γ be the set of all points in Ω such that $\dot{V}_1(z) = 0$. Let M be the largest invariant set in Γ. LaSalle's theorem ensures that every solution starting in Ω approaches M as $t \to \infty$. Let us now compute the largest invariant set M in Γ.

In the set Γ, $\dot{V}_1 = 0$ and from (7.25) it follows that $\tau = 0$ in Γ. Note that $\dot{V}_1 = 0$ also at the stable equilibrium point $(q_1, \dot{q}_1, \dot{q}_2) = (\pi, 0, 0)$. Recall that the pendulum's energy is $E(\pi, 0, 0) = -2\bar{m}g$ at the stable equilibrium point. A way to avoid this undesired convergence point is to constrain the initial conditions. Indeed, if the initial state is such that

$$V_1(0) < 2k_E \bar{m}^2 g^2 \qquad (7.26)$$

then, in view of (7.23) and given that $V_1 \leq V_1(0)$, the energy E will never reach the value $-2\bar{m}g$, which characterizes the stable equilibrium point $(q_1, \dot{q}_1, \dot{q}_2) = (\pi, 0, 0)$. The inequality (7.26) mainly imposes upper bounds on $|\dot{q}_1|$ and $|\dot{q}_2|$.

Since $\dot{E} = \dot{q}_2 \tau$ (see (7.12)) and $\tau = 0$ in the invariant set, then E is constant. From (7.23), it follows that

$$d_{21} \dot{q}_1 + d_{22} \dot{q}_2 = K \qquad (7.27)$$

for some constant K. Then, from (7.24), we get $E\dot{q}_2 = -\frac{k_v K}{k_E}$. We will consider two cases: a) $E = 0$ and b) $E = \bar{K} \neq 0$, for some constant $\bar{K}$.

- Case a: $E = 0$. From (7.24), we have

$$d_{21} \dot{q}_1 + d_{22} \dot{q}_2 = 0 \qquad (7.28)$$

Introducing (7.28) in (7.11), with $E = 0$, it then follows that

$$\begin{aligned} E &= \dot{q}^T D\dot{q} + \bar{m}g(\cos q_1 - 1) = 0 \\ &= \frac{\det(D)}{d_{22}} \dot{q}_1^2 + \bar{m}g(\cos q_1 - 1) = 0 \qquad (7.29) \end{aligned}$$

Since $\det(D)$ is a constant, equation (7.29) defines a particular trajectory called a homoclinic orbit of the pendulum in the $(q_1, \dot{q}_1)$ phase plane, which is a two-dimensional subspace of the four-dimensional state space of the complete system. (see [24, 59]). It means that $\dot{q}_1 = 0$ only when $q_1 = 0$. The pendulum moves clockwise or counter-clockwise until it reaches the equilibrium point $(q_1, \dot{q}_1) = (0, 0)$. Then, from (7.28), it follows that $\dot{q}_2 = 0$ also.

- Case b: $E = \bar{K} \neq 0$. Since $E\dot{q}_2$ is constant, then $\dot{q}_2$ is also constant. Thus from (7.27), $\dot{q}_1$ is also constant. From (7.7)-(7.10) and since $\tau = 0$, we obtain

$$\ddot{q}_1 = \frac{d_{22}}{\det(D)}\bar{m}g\sin q_1 \tag{7.30}$$

Since $\dot{q}_1$ is constant, we have $\ddot{q}_1 = 0$ and from (7.30), we conclude that $q_1 = 0[\pi]$. Note that the case when $q_1 = \pi[2\pi]$ has been excluded by imposing the constraint (7.26). Suppose now that $\dot{q}_2 \neq 0$. From (7.24), since $\tau = 0$ and $\dot{q}_1 = 0$, it follows that

$$E = -\frac{k_v d_{22}}{k_E} < 0 \tag{7.31}$$

However, since $q_1 = 0$, the energy (7.11) becomes

$$E = d_{22}\dot{q}_2^2 > 0 \tag{7.32}$$

Therefore equations (7.31) and (7.32) lead to a contradiction, which proves that the assumption $\dot{q}_2 = 0$ is false. We finally conclude that $\dot{q}_2 = 0$. Moreover, when $\dot{q}_2 = 0$, $q_1 = 0$ and $\dot{q}_1 = 0$, E is also zero, which contradicts the assumption $E \neq 0$.

Finally, the largest invariant set is given by the homoclinic orbit (7.29) together with the kinematic constraint (7.28). LaSalle's invariance principle guarantees that the system trajectories asymptotically converge to this invariant set, provided that the initial conditions are such that (7.26) is satisfied.

7.4 Second energy-based controller

In this section, we will propose another approach based on the total energy of the system, using a similar strategy developed in previous works [59] and [24]. Contrary to the algorithm proposed in the previous section, the controller presented next will be such that the wheel angular position q_2 will also converge to zero. Define the following Lyapunov function candidate

$$V_2 = \frac{1}{2}k_E E^2 + \frac{1}{2}k_v \dot{q}_2^2 + \frac{1}{2}k_p q_2^2 \tag{7.33}$$

where k_E, k_v and k_p are strictly positive constants. From (7.7)-(7.10), we get

$$\ddot{q}_2 = -\frac{d_{21}}{\det(D)}\bar{m}g\sin q_1 + \frac{d_{11}}{\det(D)}\tau \tag{7.34}$$

Therefore

$$
\begin{aligned}
\dot{V}_2 &= k_E E \dot{E} + k_v \dot{q}_2 \ddot{q}_2 + k_p \dot{q}_2 q_2 \\
&= \dot{q}_2 \left(k_E E \tau - k_1 \sin q_1 + k_2 \tau + k_p q_2 \right)
\end{aligned}
\tag{7.35}
$$

where $k_1 = \frac{k_v d_{21} \bar{m} g}{\det(D)}$; $k_2 = \frac{k_v d_{11}}{\det(D)}$. We propose a controller τ such that

$$
\tau(k_E E + k_2) - k_1 \sin q_1 + k_p q_2 = -k_d \dot{q}_2
\tag{7.36}
$$

which leads to

$$
\dot{V}_2 = -k_d \dot{q}_2^{\,2}
\tag{7.37}
$$

In view of (7.11), we have

$$
E \geq -2\bar{m}g
\tag{7.38}
$$

In order to avoid singularity in (7.36), it suffices to choose k_E and k_v such that for some $\epsilon > 0$

$$
k_E E + k_2 \geq k_E(-2\bar{m}g) + k_v \frac{d_{11}}{\det(D)} \geq \epsilon > 0
\tag{7.39}
$$

The stabilizing controller is of the form

$$
\tau = \frac{-k_d \dot{q}_2 - k_p q_2 + k_1 \sin q_1}{k_E E + k_2}
\tag{7.40}
$$

From (7.33) and (7.37), we conclude that $V_2 \leq V_2(0)$. This implies that the energy E remains bounded as well as $\dot{q}_1$, q_2 and $\dot{q}_2$. Thus the closed-loop state vector $z = [\cos q_1, \sin q_1, \dot{q}_1, q_2, \dot{q}_2]^T$ is bounded and we can thus apply LaSalle's invariance principle, as has been done in the previous section.

Defining a compact (closed and bounded) set Ω and Γ the set of all points in Ω such that $\dot{V}_2(z) = 0$. Let M be again the largest invariant set in Γ. LaSalle's theorem ensures that every solution starting in Ω approaches M as $t \to \infty$. Let us now compute the largest invariant set M in Γ.

In the set Γ, $\dot{V}_2 = 0$ and from (7.37) it follows that $\dot{q}_2 = 0$ and thus q_2 is constant in Γ. Note that $\dot{V}_2 = 0$ also at the stable equilibrium point $(q_1, \dot{q}_1, \dot{q}_2) = (\pi, 0, 0)$ and that $\tau = 0$ at this point (see (7.40)). Recall that the pendulum's energy is $E(\pi, 0, 0) = -2\bar{m}g$ at the stable equilibrium point. We will avoid this undesired convergence point by imposing a constraint on the initial conditions. Indeed, if the initial state is such that

$$V_2(0) < 2k_E \bar{m}^2 g^2 \tag{7.41}$$

then, in view of (7.33) and given that $V_2 \leq V_2(0)$, the energy E will never reach the value $-2\bar{m}g$, which characterizes the stable equilibrium point for $\dot{q}_2$. The inequality (7.41) imposes upper bounds on $|\dot{q}_1|$ and $|\dot{q}_2|$. Note that since τ in (7.40) is bounded and $\dot{q}_2 = 0$ in the invariant set Γ, then E is constant (see (7.12)). Equation (7.34) can be rewritten as

$$\ddot{q}_2 = -\frac{k_1}{k_v} \sin q_1 + \frac{k_2}{k_v} \tau \tag{7.42}$$

Therefore, since $\dot{q}_2 = 0$ we have

$$k_2 \tau = k_1 \sin q_1 \tag{7.43}$$

Introducing the above in (7.36) and since $\dot{q}_2 = 0$, we obtain

$$k_E E \tau + k_p q_2 = 0 \tag{7.44}$$

We conclude that $E\tau$ is constant in Γ. Since E is also constant, we either have a) $E = 0$ or b) $E \neq 0$.

- Case a: If $E = 0$, then from (7.44) $q_2 = 0$. Note that τ in (7.40) is bounded since $|E| < 2\bar{m}g$. Moreover, in view of (7.7), (7.8) and (7.11), since $\dot{q}_2 = 0$ and $E = 0$, (7.11) reduces to

$$E = \frac{1}{2} d_{11} \dot{q}_1^2 + \bar{m}g(\cos(q_1) - 1) = 0 \tag{7.45}$$

which defines a homoclinic orbit of the pendulum in the $(q_1, \dot{q}_1)$ phase plane. In this case, we conclude that q_2, $\dot{q}_2$ and E converge to zero. Note that τ does not necessarily converge to zero.

- Case b: If $E \neq 0$ and since $E\tau$ is constant, then the control input τ is also constant. We will prove next that in this case $\tau = 0$ in Γ.

From (7.9) and (7.10), we get

$$
\begin{aligned}
d_{11} \ddot{q}_1 &= \bar{m}g \sin q_1 & (7.46) \\
d_{21} \ddot{q}_1 &= \tau & (7.47)
\end{aligned}
$$

Introducing (7.47) in (7.46), we obtain

$$\bar{m}g \sin q_1 = \frac{d_{11}}{d_{21}}\tau \qquad (7.48)$$

Differentiating (7.48), it follows

$$\bar{m}g \cos q_1 \dot{q}_1 = 0 \qquad (7.49)$$

We conclude that either $\dot{q}_1 = 0$ or $\cos(q_1) = 0$. If $\dot{q}_1 = 0$, then from (7.47) we get $\tau = 0$. If $\cos(q_1) = 0$, then q_1 is constant and we conclude also that $\tau = 0$. We therefore conclude that $\tau = 0$ in Γ. Since $\tau = 0$, then from (7.48) $\sin(q_1) = 0$. This implies that $q_1 = 0, \pm 2\pi, \dots$. Note that the points $q_1 = \pi, 3\pi\dots$ have been avoided by imposing the constraint (7.41). From (7.44) we get $q_2 = 0$.

We finally conclude that the largest invariant set is $M = \{q_2 = 0, E = 0\}$. LaSalle's invariance principle guarantees that the system trajectories asymptotically converge to this invariant set, provided that the initial conditions are such that (7.41) is satisfied.

7.5   Simulation results

In this section, we present the simulation results using MATLAB and SIMULINK. In the model, we used the real system parameters of the reaction wheel pendulum at the University of Illinois at Urbana-Champaign. $m_1 = 0.02$ kg, $m_2 = 0.3$ kg, $l_1 = 0.125$ m, $l_{c1} = 0.063$ m, $I_1 = 47 \times 10^{-6}$ kg.m^2, $I_2 = 32 \times 10^{-6}$ kg.m^2 and $g = 9.804$. The initial conditions are

$$q_1 = 0.8\pi \qquad \dot{q}_1 = 0$$

$$q_2 = 0 \qquad \dot{q}_2 = 0$$

For the first approach, we chose the gains $k_E = 0.01$, $k_v = 200$ and $k_d = 0.1$ and the results are shown in Figure 7.2. For the second approach, we chose the gains $k_E = 400$, $k_v = 0.01$, $k_p = 0.1$ and $k_d = 0.05$. Note that the gains k_E and k_v satisfy the condition (7.39). The results are shown in Figures 7.3 and 7.4.

In both cases, the Lyapunov function decreases as expected. The energy converges monotonically to zero, but note this is not necessarily

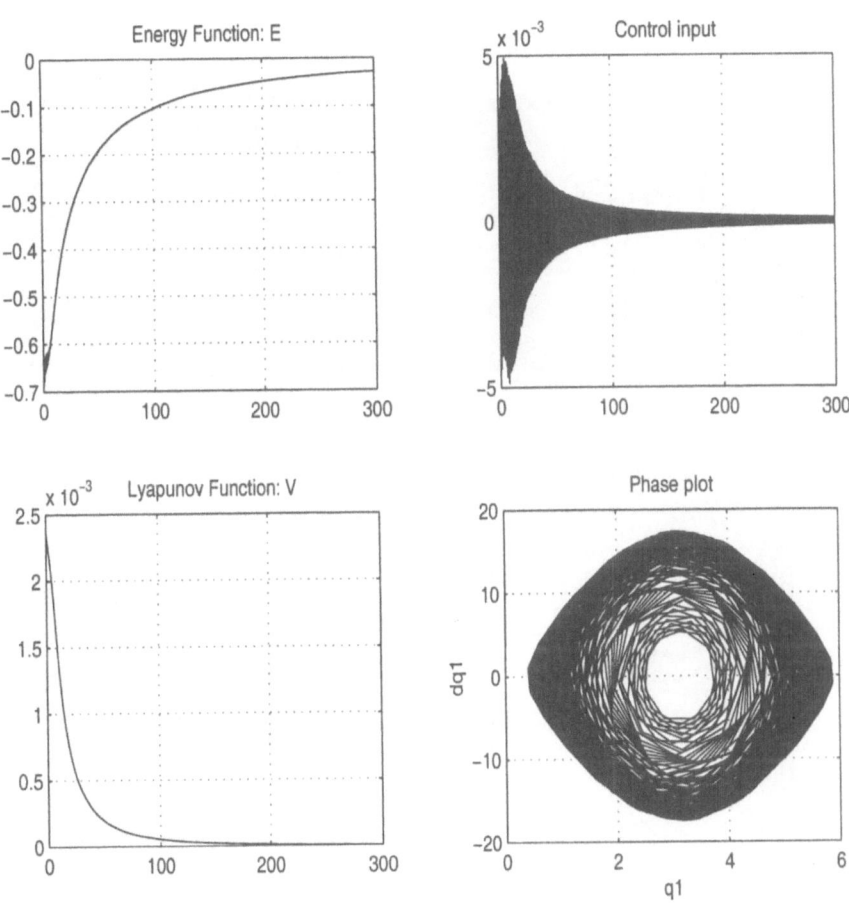

Figure 7.2: Simulation results using the first controller (7.24)

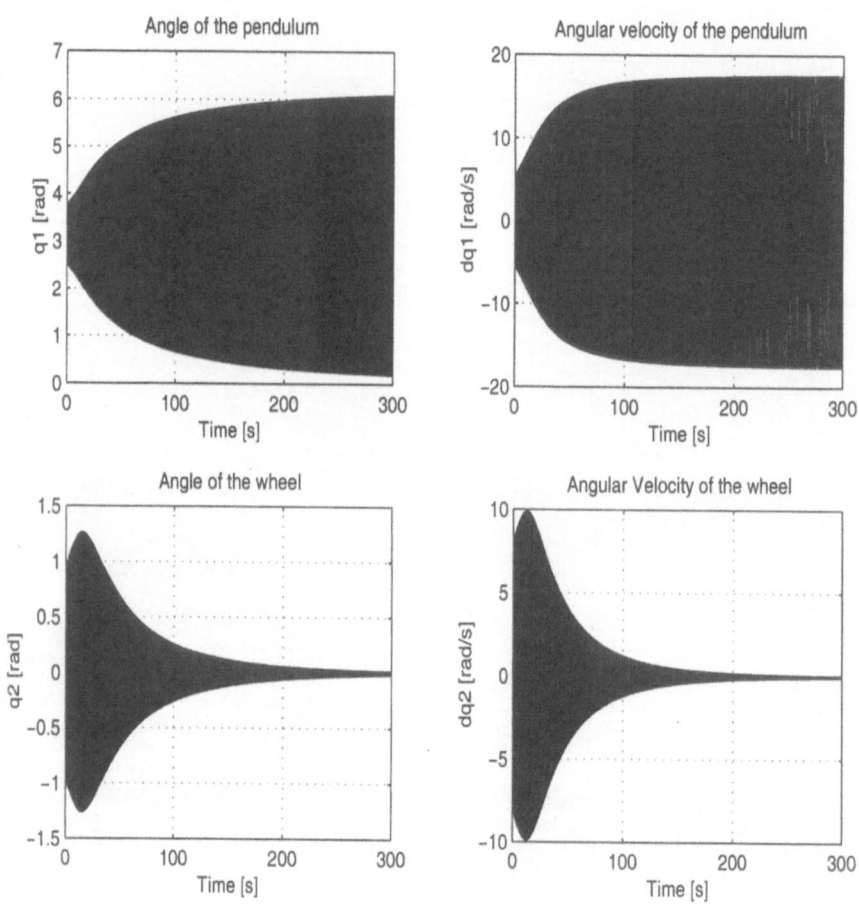

Figure 7.3: Simulation results using the second controller (7.40)

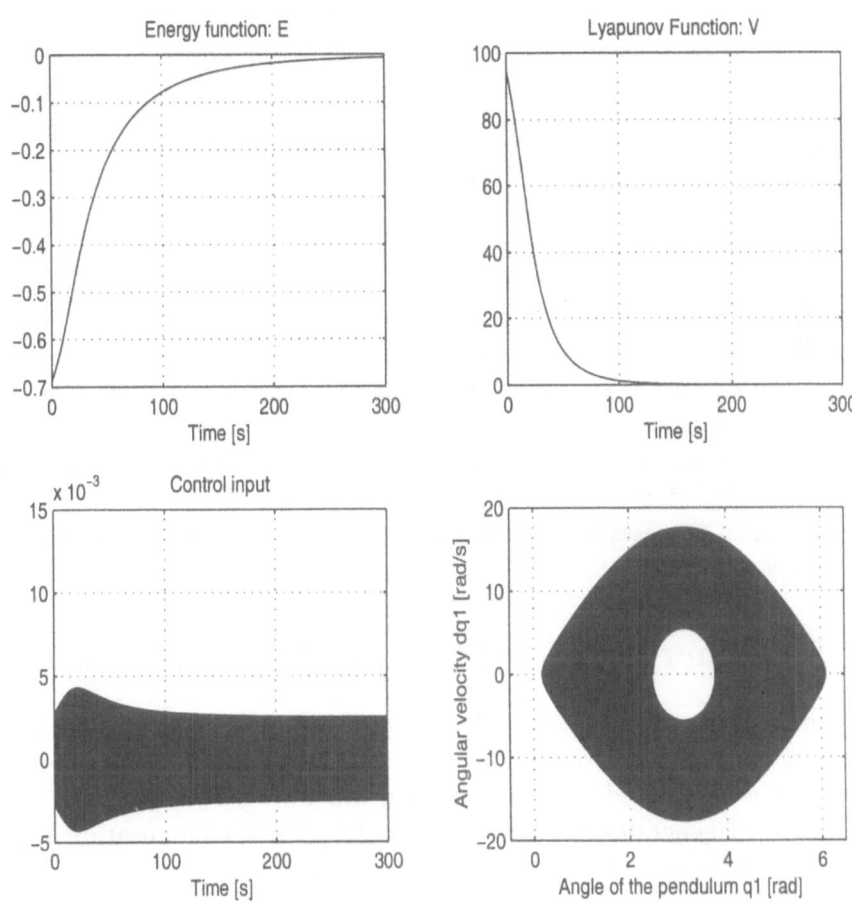

Figure 7.4: Simulation results using the second controller (7.40)

always the case. The control input τ converges to zero for the first controller (see Figure 7.2) but does not converge to zero for the second controller (see Figure 7.4). In both cases, the control input magnitude is acceptable. The phase plots show convergence to the homoclinic orbit in both cases. We have observed that the convergence rate is larger for the first controller. The closed-loop behavior strongly depends on the controller parameters k_E, k_v and k_d. The parameters used in the simulations have been selected by trying different values.

7.6   Conclusions

We have proposed two alternative approaches to swing up the reaction wheel pendulum. Both approaches are based on the total energy of the system and guarantee convergence of the pendulum to a homoclinic orbit. The first controller is such that the torque input can be saturated to any arbitrary value. The second controller is such that the wheel reaches a desired position. Simulations have shown good performance of both controllers.

We will give below a generalization of the main ideas developed for the examples of pendulum systems for a class of Euler-Lagrange systems, which possesses particular properties. It is related to the work of Shiriaev et al. [100].

7.7   Generalization for Euler-Lagrange systems

In this section, we propose to formalize the results developed for pendulum systems and give some general conditions under which the technique can be applied. The Lagrangian function L of a system having a vector of generalized variables $q^T = [x, y]$ in a configuration space $S = X \times Y$ is given by

$$L(q, \dot{q}) \;=\; K(q, \dot{q}) - P(q) \tag{7.50}$$

$$L(q, \dot{q}) \;=\; \frac{1}{2}\dot{q}^T D(q)\dot{q} - P(q) \tag{7.51}$$

where $P(q)$ is the potential energy of the system. The corresponding equations of motion are derived using Euler-Lagrange's equations

$$\frac{d}{dt}\nabla_{\dot{x}}L - \nabla_x L \;=\; \tau \tag{7.52}$$

$$\frac{d}{dt}\nabla_{\dot{y}}L - \nabla_y L \;=\; 0 \tag{7.53}$$

The control input vector is given by $U = [\tau, 0]^T$.

The equations of motion of the system can also be written in standard form, as follows

$$D(q)\ddot{q} + C(q, \dot{q})\dot{q} + G(q) = U \tag{7.54}$$

with the following properties: $D(q)$ is symmetric, positive definite and $\dot{D} - 2C$ is a skew-symmetric matrix. Moreover, P is related to $G(q)$ as follows

$$G(q) = \nabla_q P \tag{7.55}$$

The total energy of the system is given by

$$\begin{aligned} E(q, \dot{q}) &= K(q, \dot{q}) + P(q) & (7.56) \\ &= \frac{1}{2}\dot{q}^T D(q)\dot{q} + P(q) & (7.57) \end{aligned}$$

Then, using (7.54), (7.55), (7.52) and (7.53), the time derivative of E is given by

$$\frac{d}{dt}E(q(t), \dot{q}(t)) = \dot{x}(t)^T \tau(t) \tag{7.58}$$

Therefore, the total energy satisfies the passivity property. Let us consider the desired energy E_d and the desired vector x_d.

We propose the following Lyapunov function candidate

$$V(q, \dot{q}) = \frac{k_E}{2}(E(q, \dot{q}) - E_d)^2 + \frac{k_v}{2}|\dot{x}|^2 + \frac{k_x}{2}|x - x_d|^2 \tag{7.59}$$

Differentiating V along the solutions of system (7.54), we obtain

$$\dot{V} = \dot{x}^T\left[k_E(E - E_d)\tau + k_v\ddot{x} + k_x(x - x_d)\right] \tag{7.60}$$

$$\dot{V} = \dot{x}^T\left[\left(k_E(E - E_d)I + k_v[I\ 0]D(q)^{-1}\begin{bmatrix} I \\ 0 \end{bmatrix}\right)\tau + F(q, \dot{q})\right] \tag{7.61}$$

where $F(q, \dot{q})$ is a function that depends on L, x_d, the parameters k_E, k_v, k_x and is given by

$$F(q, \dot{q}) = k_v[I\ 0]D(q)^{-1}[-C\dot{q} - G] + k_x(x - x_d) \tag{7.62}$$

We propose a control law such that

$$\left(k_E(E - E_d)I + k_v[I\ 0]D(q)^{-1}\begin{bmatrix} I \\ 0 \end{bmatrix} \right)\tau + F(q,\dot{q}) = -k\dot{x} \qquad (7.63)$$

which will lead to

$$\dot{V} = -k\dot{x}^T\dot{x} \le 0 \qquad (7.64)$$

if the following matrix is invertible

$$k_E(E - E_d)I + k_v[I\ 0]D(q)^{-1}\begin{bmatrix} I \\ 0 \end{bmatrix} \qquad (7.65)$$

We assume that the energy function is bounded from below, which is normally the case for the pendulums. Therefore, there exist some positive parameters k_E and k_v such that the matrix (7.65) is strictly positive definite and thus invertible. Then, the stability analysis will be based on LaSalle's invariance principle. The objective is to prove that along the closed-loop system solutions $(q(t), \dot{q}(t))$, $\lim_{t \to +\infty} E = E_d$ and $\lim_{t \to +\infty} x = x_d$. On the other hand, this part is in general not straightforward and has actually only been developed for each particular system. Further studies on the subject are underway.

Chapter 8

The planar flexible-joint robot

8.1 Introduction

In the two previous systems, i.e. the inverted pendulum in Chapter 3 and the pendubot in Chapter 5, gravitational forces are present in both systems, since the pendulum and the pendubot move in a vertical plane. On the other hand, a pendulum or a pendubot moving in a horizontal plane has no potential energy. No gravitational forces are applied on the system.

Numerous authors are also interested in controlling planar underactuated manipulators with completely free joints. This is a challenging non-linear control problem because it requires full exploitation of the non-holonomic properties. Nakamura et al. [74] investigated the non-linear behavior of a two-joint planar manipulator with the second joint left free, from a non-linear dynamics point of view. They proposed a positioning control for both links using a time-periodic input and presented an amplitude modulation of the feedback error. Arai et al. [2] considered non-holonomic control of a three-DOF planar underactuated manipulator, with the third joint being passive. They also proved controllability of the system by constructing examples of the input trajectories from arbitrary initial states to arbitrary desired states. The trajectories for positioning are composed of simple translational and rotational trajectories segments that are stabilized by non-linear feedback control. Note that the robot arms studied in [74] and [2] are really underactuated and non controllable in their linear approximation.

In the present chapter, we will first consider a two-link planar robot with a spring between the links and only one actuator acting on the first link. The idea of considering such a system with a spring will enable us to develop an approach based on the total energy. The motivation for such a system is that a single-link flexible robot can be approximated by an n-rigid link robot with springs between the links. The practical relevance of controlling such a system comes from the need of having lighter robot arms, in particular for space applications.

Our control objective is to bring the two-link system to the origin. The passivity properties of the two-link robotic mechanism are used as a guideline in the controller synthesis, as has been done in the works [21, 24, 59] and in Chapters 3 and 5. In the present chapter, it turns out that we can propose a simpler controller compared to the ones developed in Chapter 3, Chapter 5 and [21]. The control algorithm as well as the convergence analysis is based on Lyapunov theory. Then, we will apply the same approach to a three-link robot and stabilize the system at the origin.

In Section 8.2, we recall the equations of motion of a two-link robot with a spring between the two links and its properties. Then, in Section 8.3, we will propose a control law based on an energy approach. In Section 8.4, we will analyze the stability properties of the closed-loop system. The performance of the proposed control law is shown in a simulation example in Section 8.5. In Section 8.6, the dynamic equations of a three-link robot are given. Then, in Section 8.7 we will use the same idea as in Section 8.3 to control the three-link robot. The performance of the proposed control law will be again illustrated by means of simulations in Section 8.9. Section 8.10 gives some conclusions and remarks.

8.2   The two-link planar robot

8.2.1   Equations of motion

Consider the two-link underactuated planar robot, with a spring between the two links (see Figure 8.1). We will assume that the system moves on a horizontal plane ($0xy$).

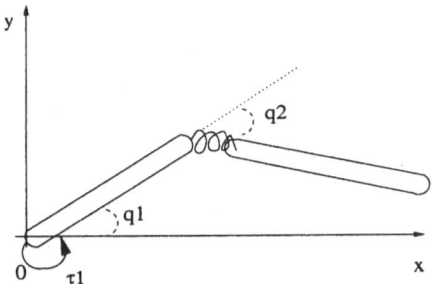

Figure 8.1: The two-link robot system

m_1	:	Mass of link 1
m_2	:	Mass of link 2
l_1	:	Length of link 1
l_2	:	Length of link 2
l_{c_1}	:	Distance to the center of mass of link 1
l_{c_2}	:	Distance to the center of mass of link 2
I_1	:	Moment of inertia of link 1 about its centroid
I_2	:	Moment of inertia of link 2 about its centroid
g	:	Acceleration due to gravity
q_1	:	Angle that link 1 makes with axis (Ox)
q_2	:	Angle that link 2 makes with link 1
τ_1	:	Torque applied on link 1

The three useful parameters are

$$\left\{ \begin{array}{ll} \theta_1 & = m_1 l_{c_1}^2 + m_2 l_1^2 + I_1 \\ \theta_2 & = m_2 l_{c_2}^2 + I_2 \\ \theta_3 & = m_2 l_1 l_{c_2} \end{array} \right. \tag{8.1}$$

Using the Euler-Lagrange formulation, the dynamic equations of the system are

$$D(q)\ddot{q} + C(q, \dot{q})\dot{q} + Kq = \tau \tag{8.2}$$

where $q = \begin{bmatrix} q_1 \\ q_2 \end{bmatrix}$ is the vector of generalized coordinates, $\tau = \begin{bmatrix} \tau_1 \\ 0 \end{bmatrix}$ is the vector of joint torques, $D(q)$ is the inertia matrix, $C(q, \dot{q})$ is the vector of Coriolis and centrifugal forces and $K = \begin{bmatrix} 0 & 0 \\ 0 & k \end{bmatrix}$ is the stiffness matrix of the spring.

$$D(q) = \begin{bmatrix} \theta_1 + \theta_2 + 2\theta_3 \cos q_2 & \theta_2 + \theta_3 \cos q_2 \\ \theta_2 + \theta_3 \cos q_2 & \theta_2 \end{bmatrix} \qquad (8.3)$$

$$C(q, \dot{q}) = \begin{bmatrix} -\theta_3 \sin(q_2)\, \dot{q}_2 & -\theta_3 \sin(q_2)\, \dot{q}_2 - \theta_3 \sin(q_2)\, \dot{q}_1 \\ \theta_3 \sin(q_2)\, \dot{q}_1 & 0 \end{bmatrix} \qquad (8.4)$$

The matrix $D(q)$ is symmetric, positive definite for all q. Moreover, $\dot{D}(q) - 2C(q, \dot{q})$ is a skew-symmetric matrix. This important property of skew-symmetric matrices will be used in establishing the passivity property of the system.

8.2.2   Linearization of the system

In order to study the controllabilty properties of the system, we will linearize its equations of motion about the origin. Let us first recall that the dynamic equations of the system can be rewritten by (see (8.2), (8.3) and (8.4))

$$\ddot{q}_1 = \frac{1}{\theta_1 \theta_2 - \theta_3^2 \cos^2 q_2} \Big[\theta_2 \theta_3 \sin q_2 \, (\dot{q}_1 + \dot{q}_2)^2 + \theta_3^2 \cos q_2 \sin(q_2)\, \dot{q}_1^2 \\ + (\theta_2 + \theta_3 \cos q_2) k q_2 + \theta_2 \tau_1 \Big] \qquad (8.5)$$

$$\ddot{q}_2 = \frac{1}{\theta_1 \theta_2 - \theta_3^2 \cos^2 q_2} \Big[-\theta_3 (\theta_2 + \theta_3 \cos q_2) \sin q_2 \, (\dot{q}_1 + \dot{q}_2)^2 \\ - (\theta_1 + \theta_3 \cos q_2) \sin(q_2)\, \dot{q}_1^2 - (\theta_2 + \theta_3 \cos q_2) \tau_1 \\ - (\theta_1 + \theta_2 + 2\theta_3 \cos q_2) k q_2 \Big] \qquad (8.6)$$

Let us consider the vector state $Y = [q_1, \dot{q}_1, q_2, \dot{q}_2]$. Differentiating equations (8.5) and (8.6) with respect to the states and evaluating them at the origin leads to the following linear system

$$\frac{d}{dt} \begin{bmatrix} q_1 \\ \dot{q}_1 \\ q_2 \\ \dot{q}_2 \end{bmatrix} = \begin{bmatrix} 0 & 1 & 0 & 0 \\ 0 & 0 & \frac{k(\theta_2 + \theta_3)}{\theta_1 \theta_2 - \theta_3^2} & 0 \\ 0 & 0 & 0 & 1 \\ 0 & 0 & \frac{-k(\theta_1 + \theta_2 + 2\theta_3)}{\theta_1 \theta_2 - \theta_3^2} & 0 \end{bmatrix} \begin{bmatrix} q_1 \\ \dot{q}_1 \\ q_2 \\ \dot{q}_2 \end{bmatrix} + \begin{bmatrix} 0 \\ \frac{\theta_2}{\theta_1 \theta_2 - \theta_3^2} \\ 0 \\ \frac{-\theta_2 - \theta_3}{\theta_1 \theta_2 - \theta_3^2} \end{bmatrix} \tau_1$$

$$= AY + B\tau_1$$

We then have

$$B = \begin{bmatrix} 0 \\ \frac{\theta_2}{\theta_1\theta_2-\theta_3^2} \\ 0 \\ \frac{-\theta_2-\theta_3}{\theta_1\theta_2-\theta_3^2} \end{bmatrix} \qquad AB = \begin{bmatrix} \frac{\theta_2}{\theta_1\theta_2-\theta_3^2} \\ 0 \\ \frac{-\theta_2-\theta_3}{\theta_1\theta_2-\theta_3^2} \\ 0 \end{bmatrix}$$

$$A^2B = \begin{bmatrix} 0 \\ \frac{-k(\theta_2+\theta_3)^2}{\theta_1\theta_2-\theta_3^2} \\ 0 \\ \frac{k(\theta_2+\theta_3)(\theta_1+\theta_2+2\theta_3)}{\theta_1\theta_2-\theta_3^2} \end{bmatrix} \qquad A^3B = \begin{bmatrix} \frac{-k(\theta_2+\theta_3)^2}{\theta_1\theta_2-\theta_3^2} \\ 0 \\ \frac{k(\theta_2+\theta_3)(\theta_1+\theta_2+2\theta_3)}{\theta_1\theta_2-\theta_3^2} \\ 0 \end{bmatrix}$$

and $\det\left(B|AB|A^2B|A^3B\right) = \frac{k^2(\theta_2+\theta_3)^2}{(\theta_1\theta_2-\theta_3^2)^4}$. The linearized system is controllable. Therefore, a full state feedback controller $\tau_1 = -K^TY$ with an appropriate gain vector K is able to successfully stabilize the system around the origin.

8.2.3 Passivity of the system

The total energy of the system is given by

$$E = \tfrac{1}{2}\dot{q}^T D(q)\dot{q} + \tfrac{1}{2}kq_2^2 \tag{8.7}$$

Differentiating this function, we obtain successively

$$\dot{E} = \dot{q}^T D(q)\ddot{q} + \tfrac{1}{2}\dot{q}^T \dot{D}(q)\dot{q} + kq_2\dot{q}_2$$

$$= \dot{q}^T(-C(q,\dot{q})\dot{q} - Kq + \tau) + \tfrac{1}{2}\dot{q}^T \dot{D}(q)\dot{q} + \dot{q}^T Kq \tag{8.8}$$

$$= \dot{q}^T\tau = \dot{q}_1\tau_1$$

Integrating both sides of the above equation yields

$$\int_0^t \dot{q}_1\tau_1 dt - E\left(q(t),\dot{q}(t)\right) - E(q(0),\dot{q}(0)) \geq -E(q(0),\dot{q}(0)) \tag{8.9}$$

As a consequence, the system with τ_1 as input and $\dot{q}_1$ as output is passive. Note that for $\tau_1 = 0$, the system (8.2) has an infinity of equilibrium points, provided that $q_2 = 0$. Our control objective is to stabilize the system around $(q_1,\dot{q}_1,q_2,\dot{q}_2) = (0,0,0,0)$ in the coordinate frame.

8.3   Control law for the two-link manipulator

It follows from (8.7) that the desired value of the energy at the origin is zero. The Lyapunov function candidate proposed here is

$$V(q, \dot{q}) = k_E E(q, \dot{q}) + \frac{k_P}{2} q_1^2 \qquad (8.10)$$

where k_E and k_P are strictly positive constants. Note that $V(q, \dot{q})$ is a positive definite function for $k > 0$. Differentiating V and using (8.8), we obtain

$$\dot{V} = \dot{q}_1 (k_E \tau_1 + k_P q_1) \qquad (8.11)$$

We propose a simple PD controller such that

$$\tau_1 = \frac{1}{k_E} (-k_P q_1 - k_D \dot{q}_1) \qquad (8.12)$$

where k_D is also a strictly positive constant. It leads to

$$\dot{V} = -k_D \dot{q}_1^2 \qquad (8.13)$$

8.3.1   Equivalent closed-loop interconnection

This section gives the equivalent closed-loop interconnection of the system and refers, among others, to the work of [14, 56, 57]. Note that this interpretation is a possible way to study stability properties of the system. Indeed, we can interpret the dynamics of the system as the negative interconnection of two passive blocks and then, using the passivity theorem, we can conclude on stability.

Looking at the closed-loop system (8.2) using the PD controller (8.12), we can interpret these dynamics as the interconnection of two subsystems with respect to inputs u_1, u_2 and outputs y_1 and y_2, with $y_1 = u_2$ and $y_2 = -u_1$, and

$$\begin{cases} u_1 = -\frac{k_D}{k_E} \dot{q}_1 - \frac{k_P}{k_E} q_1 \\ y_1 = \dot{q}_1 \end{cases} \qquad (8.14)$$

This is motivated by the fact that the two-link planar robot dynamics in (8.2), with the control law in (8.12), defines a passive operator between u_1 and y_1 (see (8.9)), with state vector $\begin{pmatrix} q \\ \dot{q} \end{pmatrix}$ and dynamics

$$D(q)\ddot{q} + C(q,\dot{q})\dot{q} + Kq = \tau = u_1 \qquad (8.15)$$

Let us write the second subsystem in state space form as

$$\begin{cases} \dot{z}_1 = u_2 \\ y_2 = \frac{k_P}{k_E} z_1 + \frac{k_D}{k_E} u_2 \end{cases} \qquad (8.16)$$

with $z_1(0) = q_1(0)$. Its transfer function is given by $H(s) = \frac{k_P + k_D s}{k_E s}$. Thus, $H(s)$ is positive real (PR), see Definition 2.3, in [56]. Applying the passivity theorem, it follows that $\dot{q}_1 \in L^2(\mathbb{R}^+)$. Notice that this is a consequence of the fact that $H(s)$ defines an input strictly passive operator, see Theorem 2.3 2), in [56]. The storage functions of each subsystem are equal to $k_E E$ for the first subsystem and $\frac{k_P}{2} q_1^2$ for the second subsystem. The sum of both storage functions yields the desired Lyapunov function defined in (8.10). The interconnection is depicted in Figure 8.2.

What is important is that we can systematically associate with these dissipative subsystems some Lyapunov functions that are systematically deduced from their passivity property. This is a fundamental property of dissipative systems that one can use to calculate Lyapunov functions for them.

Note also that all the fundamental theory on dissipative systems is extensively related to the well-known Kalman-Yakubovich-Popov lemma, which is one of the key results in control theory.

8.4 Stability analysis

The stability analysis carried out will be based on LaSalle's invariance principle (see for instance [46], Theorem 3.4, page 117). In order to apply LaSalle's theorem, we are required to define a compact (closed and bounded) set Ω with the property that every solution of system (8.2) that starts in Ω remains in Ω for all future time. Recall that the Lyapunov function candidate V is positive definite. Since $V(q,\dot{q})$ in (8.10) is non-increasing along the trajectories, (see (8.13)), then q, $\dot{q}$ are bounded. Therefore, the solution of the closed-loop system remains

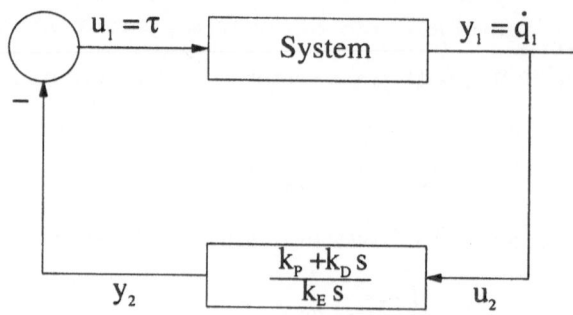

Figure 8.2: The equivalent representation as the negative feedback interconnection of a passive system and a SPR transfer function

inside a compact set Ω that is defined by the initial state values. Let Γ be the set of all points in Ω such that $\dot{V}(q, \dot{q}) = 0$. Let M be the largest invariant set in Γ. LaSalle's theorem ensures that every solution starting in Ω approaches M as $t \to \infty$. Let us now compute the largest invariant set M in Γ.

In the set Γ (see (8.13)), $\dot{V} = 0$, i.e. $\dot{q}_1 = 0$. This implies that q_1 and V are constant. From (8.10), it follows that E is also constant. Regarding (8.12), it follows that the control law has been chosen such that

$$-k_D \dot{q}_1 = k_E \tau_1 + k_P q_1 \qquad (8.17)$$

From the above equation, we conclude that τ_1 is constant in Γ. Recalling that E is also constant, we either have a) $E = 0$ or b) $E \neq 0$. Let us study both cases.

- Case a: If $E = 0$, we have

$$E = \frac{1}{2}\theta_2 \dot{q}_2^2 + \frac{1}{2}k q_2^2 = 0$$

  which implies that $q_2 = 0$ and $\dot{q}_2 = 0$. From (8.2), (8.3) and (8.4), it then follows that $\tau_1 = 0$. Therefore, from (8.17) it follows that $q_1 = 0$. The trajectories converge to the desired position.

- Case b: If $E \neq 0$ but is constant

$$E = \frac{1}{2}\theta_2\dot{q}_2^2 + \frac{1}{2}kq_2^2 = \text{cte}$$

We also have τ_1 as constant. Note that if the torque τ_1 is constant and different from zero, the position of the first link q_1 cannot remain constant. Since q_1 is constant, we conclude by contradiction that τ_1 is zero. From (8.17) it follows that $q_1 = 0$.

Note that the second link cannot move without exerting a force on the first link. Since the torque input τ_1 is zero and $q_1 = 0$, we can conclude that $\dot{q}_2$ should be zero.

All this argument can also be proved mathematically as shown below.

We will now study, in more detail, the case: $\tau_1 = cte$, $\dot{q}_1 = 0$, $\ddot{q}_1 = 0$. Our objective is now to conclude that q_2, $\dot{q}_2$, $\ddot{q}_2$ are also zero, and that $q_1 = 0$.

The system (8.2) for $\tau_1 = cte$, $\dot{q}_1 = 0$ and $\ddot{q}_1 = 0$ becomes

$$
\begin{align}
(\theta_2 + \theta_3 \cos q_2)\, \ddot{q}_2 - \theta_3 \dot{q}_2^2 \sin q_2 &= \tau_1 \tag{8.18} \\
\theta_2 \ddot{q}_2 + kq_2 &= 0 \tag{8.19}
\end{align}
$$

Introducing (8.19) into (8.18), we obtain

$$(\theta_2 + \theta_3 \cos q_2)\, kq_2 = -\theta_3\theta_2\dot{q}_2^2 \sin q_2 - \theta_2\tau_1 \tag{8.20}$$

By taking the time derivative of (8.20), we have

$$k\dot{q}_2\,(\theta_2 + \theta_3 \cos q_2) - k\theta_3 q_2\dot{q}_2 \sin q_2 = -2\theta_3\theta_2\dot{q}_2\ddot{q}_2 \sin q_2 - \theta_3\theta_2\dot{q}_2^3 \cos q_2 \tag{8.21}$$

Note that $\dot{q}_2$ is a common factor in the above equation. If $\dot{q}_2$ is zero in a given interval of time, we will conclude directly that $q_2 = 0$ (see (8.19)). If it is not zero in an interval of time, we can divide the equation above by $\dot{q}_2$.

We will therefore study separately the case when $\dot{q}_2 = 0$ and when $\dot{q}_2 \neq 0$ for some interval of time.

Case i) If $\dot{q}_2 = 0$ then $\ddot{q}_2 = 0$ and from (8.19), we have $q_2 = 0$ and actually $E = 0$.

Case ii) If $\dot{q}_2(t) \neq 0$. The equation (8.21) becomes, using (8.19)

$$
\begin{aligned}
k\left(\theta_2 + \theta_3 \cos q_2\right) - k\theta_3 q_2 \sin q_2 &= -2\theta_3 \theta_2 \ddot{q}_2 \sin q_2 - \theta_3 \theta_2 \dot{q}_2^2 \cos q_2 \\
&= 2k\theta_3 q_2 \sin q_2 - \theta_3 \theta_2 \dot{q}_2^2 \cos q_2
\end{aligned}
$$

Thus

$$
k\left(\theta_2 + \theta_3 \cos q_2\right) + \theta_3 \theta_2 \dot{q}_2^2 \cos q_2 = 3k\theta_3 q_2 \sin q_2 \qquad (8.22)
$$

Differentiating equation (8.22), we get

$$
\begin{aligned}
-k\theta_3 \dot{q}_2 \sin q_2 - \theta_2 \theta_3 \dot{q}_2^3 \sin q_2 + 2\theta_2 \theta_3 \dot{q}_2 \ddot{q}_2 \cos q_2 &= 3k\theta_3 \dot{q}_2 \left(\sin q_2 \right. \\
&\quad \left. + q_2 \cos q_2\right)
\end{aligned}
$$

Dividing the above by θ_3 and $\dot{q}_2$, and replacing $\theta_2 \ddot{q}_2$ by $-kq_2$ (see (8.19)), we obtain

$$
-\theta_2 \dot{q}_2^2 \sin q_2 = 4k \sin q_2 + 5kq_2 \cos q_2 \qquad (8.23)
$$

Taking the time derivative of (8.23), it follows that

$$
-2\theta_2 \dot{q}_2 \ddot{q}_2 \sin q_2 - \theta_2 \dot{q}_2^3 \cos q_2 = 4k\dot{q}_2 \cos q_2 - 5kq_2\dot{q}_2 \sin q_2 + 5k\dot{q}_2 \cos q_2
$$

Dividing again by $\dot{q}_2$ and replacing $\theta_2 \ddot{q}_2$ by $-kq_2$ again (see (8.19)), we obtain

$$
-\theta_2 \dot{q}_2^2 \cos q_2 = 9k \cos q_2 - 7kq_2 \sin q_2 \qquad (8.24)
$$

By taking the time derivative of (8.24), we then have

$$
-2\theta_2 \dot{q}_2 \ddot{q}_2 \cos q_2 + \theta_2 \dot{q}_2^3 \sin q_2 = -9k\dot{q}_2 \sin q_2 - 7k\dot{q}_2 \sin q_2 - 7kq_2\dot{q}_2 \cos q_2
$$

Dividing by $\dot{q}_2$, and replacing $\theta_2 \ddot{q}_2$ yields

$$
-\theta_2 \dot{q}_2^2 \sin q_2 = 16k \sin q_2 + 9kq_2 \cos q_2 \qquad (8.25)
$$

Combining (8.23) and (8.25), it follows that

$$
3 \sin q_2 + q_2 \cos q_2 = 0 \qquad (8.26)
$$

We now differentiate (8.26), to obtain

$$3\ddot{q}_2 \cos q_2 + \dot{q}_2 \cos q_2 - q_2\dot{q}_2 \sin q_2 = 0$$

Dividing by $\dot{q}_2 \neq 0$, and taking the time derivative, we obtain

$$-4\dot{q}_2 \sin q_2 - \dot{q}_2 \sin q_2 - q_2\dot{q}_2 \cos q_2 = 0$$

and so

$$5 \sin q_2 + q_2 \cos q_2 = 0 \tag{8.27}$$

From (8.26) and (8.27), we finally obtain

$$\sin q_2 = 0$$

From (8.26) and the above, we conclude that

$$q_2 = 0$$

This last conclusion contradicts the premise, i.e. $\dot{q}_2(t) \neq 0$. Therefore, the particular case ii) is not possible. The only case is finally: $q_2 = 0$, $\dot{q}_2 = 0$, $\ddot{q}_2 = 0$ and $E = 0$. So, as in case a), we finally have $\tau_1 = 0$ and from (8.17), it follows that $q_1 = 0$.

From the analysis above, the largest invariant set M is given by the desired position $(q_1, \dot{q}_1, q_2, \dot{q}_2) = (0,0,0,0)$ with $E = 0$. All the solutions converge to the invariant set M.

LaSalle´s theorem allows us to establish asymptotic stability. The convergence rate to the origin depends on the controller parameters k_P, k_D and k_E.

8.5 Simulation results

In order to observe the performance of the proposed control law based on the energy, we performed simulations on MATLAB, using SIMULINK.

We considered the system taking the parameters $\theta_{i,1 \leq i \leq 3}$ as follows: $\theta_1 = 0.0799$, $\theta_2 = 0.0244$, $\theta_3 = 0.0205$. We have some freedom in the choice of the coefficients k_P , k_D and k_E.

Figures 8.3 and 8.4 show the results for $k_P = 1$, $k_D = 1$, $k_E = 1$, for a stiffness constant $k = 1$ and for an initial position

$$q_1 = \tfrac{2\pi}{3} \qquad\qquad q_2 = \tfrac{\pi}{3}$$
$$\dot{q}_1 = 0 \qquad\qquad \dot{q}_2 = 0$$

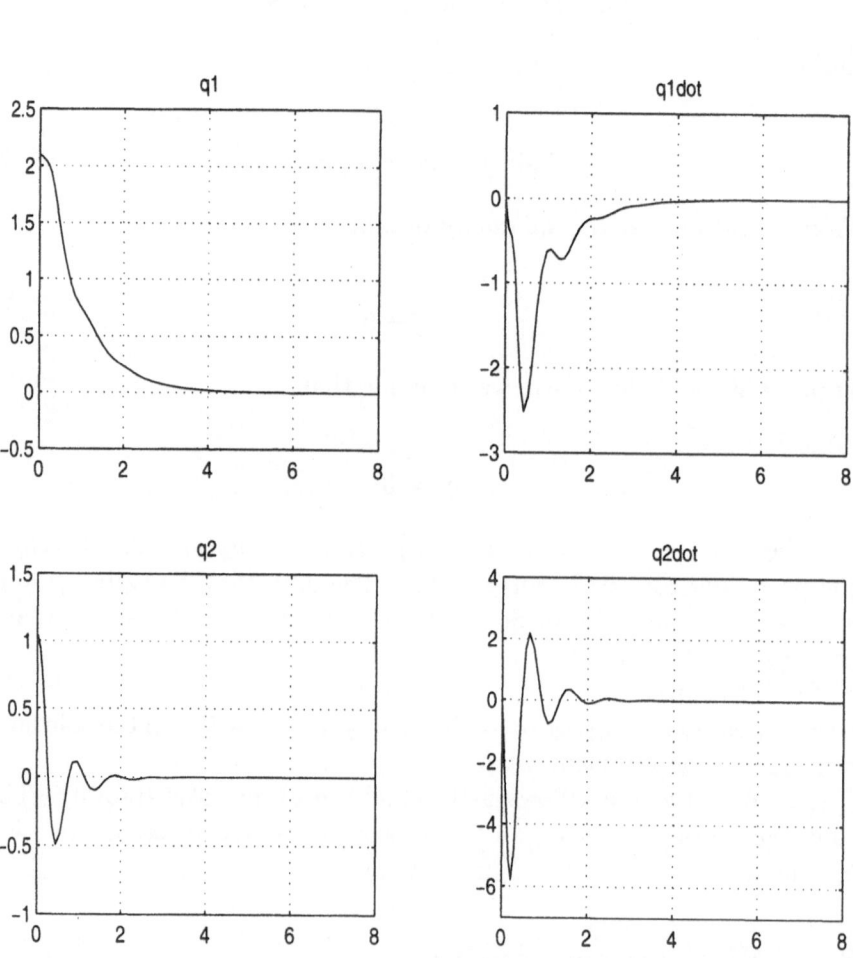

Figure 8.3: States of the system

Simulations showed that our control law brings the state of the system to the desired position. Note that the energy E goes to zero.

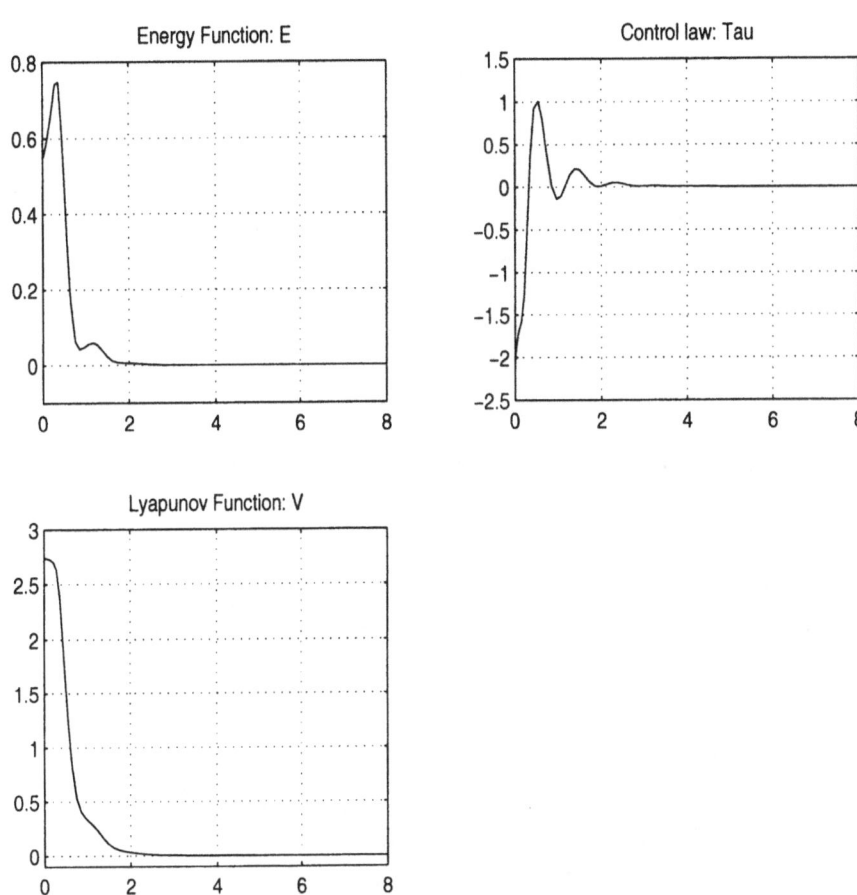

Figure 8.4: Performance of the system

8.6   The three-link planar robot

This section presents the study of an extension of the proposed control scheme for a three-link robot manipulator having an actuator only in the first link (see Figure 8.5).

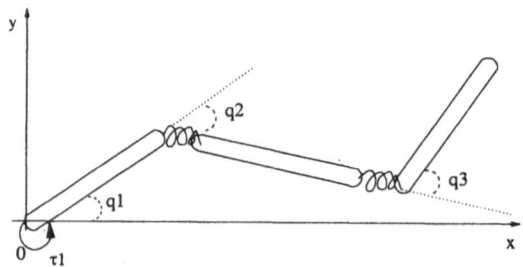

Figure 8.5: The three-link robot system

m_1	:	Mass of link 1
m_2	:	Mass of link 2
m_3	:	Mass of link 3
l_1	:	Length of link 1
l_2	:	Length of link 2
l_3	:	Length of link 3
l_{c_1}	:	Distance to the center of mass of link 1
l_{c_2}	:	Distance to the center of mass of link 2
l_{c_3}	:	Distance to the center of mass of link 3
I_1	:	Moment of inertia of link 1 about its centroid
I_2	:	Moment of inertia of link 2 about its centroid
I_3	:	Moment of inertia of link 3 about its centroid
g	:	Acceleration due to gravity
q_1	:	Angle that link 1 makes with axis(Ox)
q_2	:	Angle that link 2 makes with link 1
q_3	:	Angle that link 3 makes with link 2
τ_1	:	Torque applied on link 1

The three parameters $\theta_{i,1\leq i\leq 3}$ are the same as seen in Section 8.2.1.

The six other ones are given by

$$\begin{cases} \theta_6 &= I_3 + m_3 \left(l_{c3}^2 + l_1^2 + l_2^2\right) \\ \theta_7 &= m_3 l_1 l_{c3} \\ \theta_8 &= m_3 l_2 l_{c3} \\ \theta_9 &= m_3 l_1 l_2 \\ \theta_{10} &= I_3 + m_3 \left(l_{c3}^2 + l_2^2\right) \\ \theta_{11} &= I_3 + m_3 l_{c3}^2 \end{cases} \tag{8.28}$$

We obtain the dynamic equations of motion by using an Euler-Lagrange formulation. In matrix form, the third order system becomes

$$D(q)\ddot{q} + C(q,\dot{q})\dot{q} + Kq = \tau \tag{8.29}$$

where $q = \begin{bmatrix} q_1 \\ q_2 \\ q_3 \end{bmatrix}$ is the vector of generalized coordinates, $\tau = \begin{bmatrix} \tau_1 \\ 0 \\ 0 \end{bmatrix}$ is the vector of joint torques, $D(q)$ is the inertia matrix, $C(q,\dot{q})$ is the vector of Coriolis and centrifugal forces and $K = \begin{bmatrix} 0 & 0 & 0 \\ 0 & k_2 & 0 \\ 0 & 0 & k_3 \end{bmatrix}$ is the stiffness matrix of the two springs. Here

$$D(q) = \begin{bmatrix} d_{11} & d_{12} & d_{13} \\ d_{21} & d_{22} & d_{23} \\ d_{31} & d_{32} & d_{33} \end{bmatrix} \tag{8.30}$$

and

$$C(q,\dot{q}) = \begin{bmatrix} h\dot{q}_2 + j\dot{q}_3 & h\left(\dot{q}_1 + \dot{q}_2\right) + j\dot{q}_3 & j\left(\dot{q}_1 + \dot{q}_2 + \dot{q}_3\right) \\ -h\dot{q}_2 + h_4\dot{q}_3 & h_4\dot{q}_3 & h_4\left(\dot{q}_1 + \dot{q}_2 + \dot{q}_3\right) \\ -j\dot{q}_1 - h_4\dot{q}_2 & -h_4\left(\dot{q}_1 + \dot{q}_2\right) & 0 \end{bmatrix} \tag{8.31}$$

with

$$\begin{aligned} d_{11} =\ & \theta_1 + \theta_2 + \theta_6 + 2\left(\theta_3 + \theta_9\right)\cos q_2 \\ & + 2\theta_7 \cos\left(q_2 + q_3\right) + 2\theta_8 \cos q_3 \\ d_{12} =\ & d_{21} = \theta_2 + \theta_{10} + \left(\theta_3 + \theta_9\right)\cos q_2 \\ & + \theta_7 \cos\left(q_2 + q_3\right) + 2\theta_8 \cos q_3 \\ d_{13} =\ & d_{31} = \theta_{11} + \theta_7 \cos\left(q_2 + q_3\right) \\ & + \theta_8 \cos q_3 \\ d_{22} =\ & \theta_2 + \theta_{10} + 2\theta_8 \cos q_3 \\ d_{23} =\ & d_{32} = \theta_{11} + \theta_8 \cos q_3 \\ d_{33} =\ & \theta_{11} \end{aligned}$$

$$\begin{aligned} h =\ & h_1 + h_2 + h_3 \\ j =\ & h_3 + h_4 \\ h_1 =\ & -\theta_3 \sin q_2 \\ h_2 =\ & -\theta_9 \sin q_2 \\ h_3 =\ & -\theta_7 \sin\left(q_2 + q_3\right) \\ h_4 =\ & -\theta_8 \sin q_3 \end{aligned}$$

Recall that the following properties are satisfied, i.e. the matrix $D(q)$ is symmetric, positive definite for all q and $\dot{D}(q) - 2C(q, \dot{q})$ is a skew-symmetric matrix.

8.7   Control law for the three-link robot

Our control objective is to stabilize the system around the origin, i.e. $(q_1, \dot{q}_1, q_2, \dot{q}_2, q_3, \dot{q}_3) = (0, 0, 0, 0, 0, 0)$ in the coordinate frame. The Lyapunov function candidate proposed here is the same as for the two-link robot

$$V(q, \dot{q}) = k_E E(q, \dot{q}) + \frac{k_P}{2} q_1^2 \tag{8.32}$$

where k_E and k_P are strictly positive constants. The total energy is

$$E = \tfrac{1}{2} \dot{q}^T D(q) \dot{q} + \tfrac{1}{2} k_2 q_2^2 + \tfrac{1}{2} k_3 q_3^2 \tag{8.33}$$

and the time derivative of E is again

$$\dot{E} = \dot{q}_1 \tau_1 \tag{8.34}$$

Note that $V(q, \dot{q})$ is a positive definite function for k_2 and $k_3 > 0$. Differentiating V and using (8.34), we obtain

$$\dot{V} = \dot{q}_1 (k_E \tau_1 + k_P q_1) \tag{8.35}$$

As an immediate consequence, we propose the same controller as for two links

$$\tau_1 = \frac{1}{k_E} \left(-k_P q_1 - k_D \dot{q}_1 \right) \tag{8.36}$$

which simplifies as

$$\dot{V} = -k_D \dot{q}_1^2 \tag{8.37}$$

8.8 Stability analysis

The analysis will again be based on LaSalle's invariance principle (see Section 8.4). We also compute the largest invariant set M in Γ.

In the set Γ (see (8.37)), $\dot{V} = 0$, i.e. $\dot{q}_1 = 0$. This implies that q_1 and V are constant in the set Γ. From (8.32), it follows that E is also constant. Note that q, $\dot{q}$ are also bounded. From (8.36), it follows that the control law has been chosen such that

$$-k_D \dot{q}_1 = k_E \tau_1 + k_P q_1 \qquad (8.38)$$

From the above equation, we conclude that τ_1 is constant in Γ. Since E is constant, we either have a) $E = 0$ or b) $E \neq 0$.

- Case a: If $E = 0$, then since D is symmetric and positive definite, from (8.33) it follows that $q_2 = 0$, $\dot{q}_2 = 0$, $q_3 = 0$ and $\dot{q}_3 = 0$. From (8.29), (8.30) and (8.31) it follows that $\tau_1 = 0$. Therefore, from (8.38) it follows that $q_1 = 0$. Finally, the trajectories converge to the desired position.

- Case b: If $E \neq 0$ but is constant, from (8.33), E has the form

$$E = \frac{1}{2}\alpha_2 \dot{q}_2^2 + \frac{1}{2}\alpha_3 \dot{q}_3^2 + \sum_{i,j=2,3;i\neq j} \frac{1}{2}\beta_{ij}\dot{q}_i\dot{q}_j + \frac{1}{2}k_2 q_2^2 + \frac{1}{2}k_3 q_3^2 = \text{cte}$$

We recall that τ_1 is constant (see (8.38)). As seen in Section 8.4, if the torque τ_1 is constant and different from zero, the position of the first link q_1 cannot remain constant. Since q_1 is constant, we conclude by contradiction that τ_1 is zero. From (8.38), it follows that $q_1 = 0$. Note that the second link cannot move without exerting a force on the first link. Since the torque input τ_1 is zero and $q_1 = 0$, we can conclude that $\dot{q}_2$ should be zero. Since $\dot{q}_2 = 0$, then $\ddot{q}_2 = 0$.

The system (8.29) for $\tau_1 = 0$, $\dot{q}_1 = 0$, $\ddot{q}_1 = 0$, $q_1 = 0$, $\dot{q}_2 = 0$ and $\ddot{q}_2 = 0$ becomes

$$\begin{aligned}
0 &= \left(\theta_{11} + \theta_7 \cos(q_2 + q_3) + \theta_8 \cos q_3\right)\ddot{q}_3 \\
&\quad - \left(\theta_7 \sin(q_2 + q_3) + \theta_8 \sin q_3\right)\dot{q}_3^2 \qquad (8.39) \\
0 &= \left(\theta_{11} + \theta_8 \cos q_3\right)\ddot{q}_3 - \theta_8 \sin q_3 \dot{q}_3^2 + k_2 q_2 \qquad (8.40) \\
0 &= \theta_{11}\ddot{q}_3 + k_3 q_3 \qquad (8.41)
\end{aligned}$$

Introducing (8.41) into (8.39) and (8.40), we obtain successively

$$-\theta_8 \sin q_3 \dot{q}_3^2 = \frac{k_3 q_3}{\theta_{11}}(\theta_{11} + \theta_7 \cos(q_2 + q_3) + \theta_8 \cos q_3)$$
$$+\theta_7 \sin(q_2 + q_3)\dot{q}_3^2 \qquad (8.42)$$
$$k_2 q_2 = \frac{-k_3 \theta_7}{\theta_{11}} \cos(q_2 + q_3)q_3 - \theta_7 \sin(q_2 + q_3)\dot{q}_3^2 \quad (8.43)$$

Differentiating equation (8.43) and using $\dot{q}_2 = 0$, we get

$$0 = \frac{k_3 \theta_7}{\theta_{11}} [\sin(q_2 + q_3)\dot{q}_3 q_3 - \cos(q_2 + q_3)\dot{q}_3]$$
$$-\theta_7 \cos(q_2 + q_3)\dot{q}_3^3 - 2\theta_7 \sin(q_2 + q_3)\dot{q}_3 \ddot{q}_3 \quad (8.44)$$

Case i) If $\dot{q}_3 = 0$ then $\ddot{q}_3 = 0$. From (8.40), we have $q_2 = 0$ and from (8.41), $q_3 = 0$. Therefore, we also have $E = 0$.

Case ii) If $\dot{q}_3(t) \neq 0$

Dividing equation (8.44) by $\dot{q}_3$ and replacing $\ddot{q}_3$ by $-\frac{k_3 q_3}{\theta_{11}}$ (see (8.41)), we obtain

$$\frac{3k_3}{\theta_{11}} \sin(q_2 + q_3)q_3 - \cos(q_2 + q_3)(\dot{q}_3^2 + \frac{k_3}{\theta_{11}}) = 0 \qquad (8.45)$$

Taking the time derivative of (8.45), using $\dot{q}_2 = 0$, dividing by $\dot{q}_3$ and again replacing $\ddot{q}_3$ by $-\frac{k_3 q_3}{\theta_{11}}$, it follows that

$$\frac{5k_3}{\theta_{11}} \cos(q_2 + q_3)q_3 + \sin(q_2 + q_3)(\frac{4k_3}{\theta_1 1} + \dot{q}_3^2) = 0 \qquad (8.46)$$

Then, differentiating equation (8.46) and using the same simplifications as above, we finally obtain

$$\frac{7k_3}{\theta_{11}} \sin(q_2 + q_3)q_3 - \cos(q_2 + q_3)(\dot{q}_3^2 + \frac{9k_3}{\theta_{11}}) = 0 \qquad (8.47)$$

Introducing (8.45) into (8.47), we have

$$\frac{-7}{3} \cos(q_2 + q_3)(\dot{q}_3^2 + \frac{k_3}{\theta_{11}}) + \cos(q_2 + q_3)(\dot{q}_3^2 + \frac{9k_3}{\theta_{11}}) = 0 \quad (8.48)$$

Thus we have either case a)

$$\cos(q_2 + q_3) = 0 \qquad (8.49)$$

or case b)

$$-\dot{q_3}^2 + \frac{5k_3}{\theta_{11}} = 0 \qquad (8.50)$$

- Case a: Since $\dot{q_2} = 0$, then q_2 is constant. Using equation (8.49), q_3 is also constant. Therefore, $\dot{q_3} = 0$ and $\ddot{q_3} = 0$ too. Equation (8.41) implies $q_3 = 0$ and equation (8.40) implies $q_2 = 0$. On the other hand, this contradicts (8.49). Therefore, we only can have (8.50).

- Case b: $\dot{q_3}^2$ is constant, then $\ddot{q_3} = 0$ and with (8.41), $q_3 = 0$ and so $\dot{q_3} = 0$. This last conclusion contradicts the premise of case ii), i.e. $\dot{q_3} \neq 0$.

In conclusion, the only case is the following: $q_2 = 0$, $q_3 = 0$, $\dot{q_3} = 0$, $\ddot{q_3} = 0$ and $E = 0$ (case a)). Finally, the largest invariant set M is given by the desired position $(q_1, \dot{q_1}, q_2, \dot{q_2}, q_3, \dot{q_3}) = (0, 0, 0, 0, 0, 0)$ with $E = 0$. All the solutions converge to the invariant set.

8.9 Simulation results

The performance of the control law can be viewed on the following figures, performed by simulations.

The values of the parameters $\theta_{i,1 \leq i \leq 3}$ are the same as those taken in Section 8.5. The other ones were chosen as follows: $\theta_6 = 0.08$, $\theta_7 = 0.01$, $\theta_8 = 0.01$, $\theta_9 = 0.01$, $\theta_{10} = 0.07$, $\theta_{11} = 0.02$.

Moreover, Figures 8.6 and 8.7 give the results for $k_P = 1$, $k_D = 1$, $k_E = 1$, for stiffness constants $k_2 = 1$, $k_3 = 1$ and for an initial position

$$q_1 = \frac{2\pi}{3} \qquad\qquad q_2 = \frac{\pi}{4} \qquad\qquad q_3 = \frac{\pi}{6}$$
$$\dot{q_1} = 0 \qquad\qquad\qquad \dot{q_2} = 0 \qquad\qquad\qquad \dot{q_3} = 0$$

Simulations showed that our control law brings the state of the system to the origin. Note that the energy E goes to zero.

8.10 Conclusions

We have proposed a new control strategy for underactuated flexible-joint robot manipulators. The controller design is based on an energy approach and the stability has been studied using LaSalle's invariance

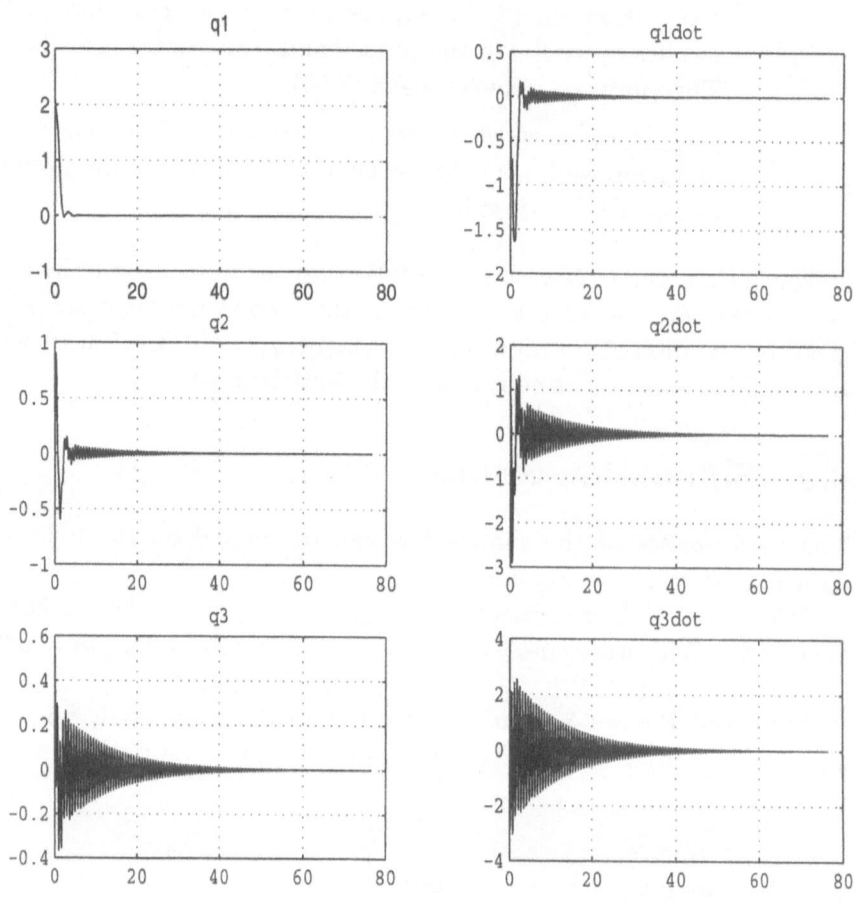

Figure 8.6: States of the system

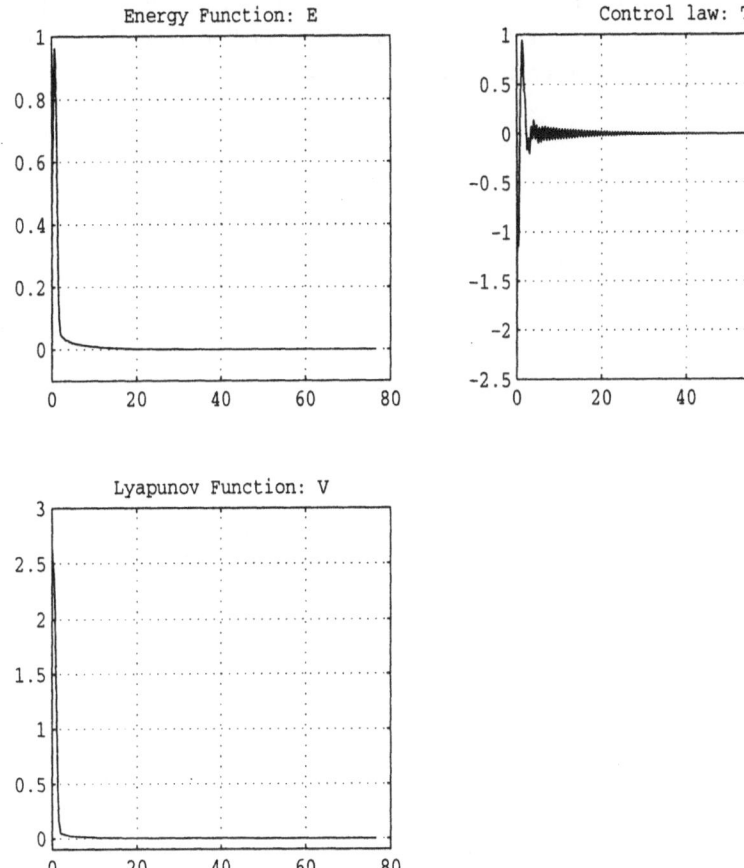

Figure 8.7: Performance of the system

principle. It has been proved that the controller globally stabilizes the origin. The technique has been developed for manipulators having two and three links and can be extended for the general case. Note that a one-flexible link manipulator can be approximated as an n-rigid link robot with springs between the links having only a motor acting on the first link. Therefore, the proposed method can be seen as an approximate way of controlling a one-flexible link robot manipulator. Examples have been presented showing that the controller performs well in simulation for the two- and three-link cases.

Chapter 9

The PPR planar manipulator

9.1 Introduction

In the last few years, we have witnessed great interest in the control of robot manipulators with elastic joints and links. In this chapter, we will consider a planar elastic manipulator with three actuated joints, wherein the third actuated joint is coupled to the end effector through an elastic joint. This system is in fact a planar robot with two prismatic and one revolute (PPR) joints, moving on a horizontal plane so that no gravitational forces appear in the system. The three PPR joints are actuated, while the elastic degree of freedom appearing in the system is not actuated. Therefore, the system is underactuated, since it has four degrees of freedom with only three control inputs. This example is a type of robot manipulators that exhibit features of joint elasticity and kinematic redundancy [88].

A model of this robot manipulator has been presented in the paper of Baillieul [8], in which he described several problems in planning motions for kinematically redundant robots with flexible components, wherein the goal is to exploit redundant degrees of freedom to minimize the dynamic effects of joint elasticity. Robots with elastic components can "store energy" and such energy storage enriches the set of possible modes of behavior of the mechanism that can be exploited to achieve control objectives that would not be possible with rigid-link mechanisms. The paper deals with the problems of using robot kinematics to avoid storing elastic energy. As an example, he developed point-to-point motions in the planar elastic manipulator using fifth order splines.

Reyhanoglu et al. [88] presented a theoretical framework for the dynamics and control of underactuated mechanical systems that satisfy non-integrable acceleration relations. They introduced a non-linear control system formulation and analyzed controllability and stabilizability properties, which can be useful for further research in this area. The planar PPR robot illustrates their results. This example is indeed a new control theoretical formulation that incorporates a specific design constraint, which imposes a zero torque on the elastic joint.

In the present work, we do not avoid storing elastic energy. We will in fact incorporate the elastic mode in our control stabilization problem.

The stabilization algorithm proposed here is related to the works of [18, 21, 24, 59] and the previous chapters in which the total energy of the systems is used in the control of underactuated manipulators. The passivity properties of the system will be analyzed and we will use an energy-based approach to establish the proposed control law. The control algorithm as well as the convergence analysis are based on Lyapunov theory and LaSalle's invariance theorem. The performance of the proposed control law is shown in simulation.

9.2   System dynamics

Consider the model of the planar (PPR) redundant manipulator with one elastic degree of freedom as presented in Figure 9.1. It consists of a hub mass body M, which can translate freely in the plane, and can be rotated by any angle θ with respect to a fixed horizontal axis and a massless arm at the tip of which the end-effector is attached. The arm is attached to the hub by a revolute joint and a torsional spring whose neutral position is $\phi = 0$. The two degrees of freedom of the translation of the mass M as well as the angle θ through which the hub is rotated can be directly controlled. The variable ϕ measures the deviation of the mechanism's arm from the assigned value θ. Whenever the variable ϕ is displaced from zero, it induces a restoring torque $-k\phi$, where k denotes the torsional spring constant. Let (x_b, y_b) be the cartesian position of the base body and let (x, y) denote the end-effector position of the manipulator. Therefore, the kinematic relationship between the configuration variables (x_b, y_b, θ, ϕ) and the end-effector position (x, y) is given by

$$\begin{pmatrix} x \\ y \end{pmatrix} = \begin{pmatrix} x_b \\ y_b \end{pmatrix} + \begin{pmatrix} l\cos(\theta + \phi) \\ l\sin(\theta + \phi) \end{pmatrix} \tag{9.1}$$

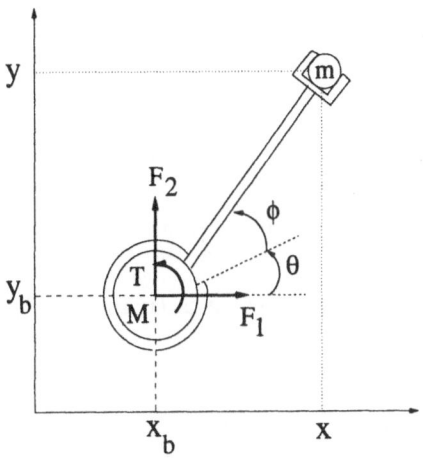

Figure 9.1: The planar PPR robot

M	:	Hub mass body
I	:	Rotational inertia
m	:	End-effector and payload combination mass
l	:	Length of massless arm
k	:	Torsional spring constant
x_b	:	Cartesian position on x axis of the hub mass body
y_b	:	Cartesian position on y axis of the hub mass body
x	:	End-effector position on x axis of the manipulator
y	:	End-effector position on y axis of the manipulator
θ	:	Angle through which the base body is rotated
ϕ	:	Deviation of the massless arm from the assigned value θ
F_1	:	Force input in the x direction applied on the base body
F_2	:	Force input in the y direction applied on the base body
T	:	Torque input applied on the base body

9.2.1   Equations of motion via Euler-Lagrange formulation

In this section, we will derive the equations of motion of the system. We first present the kinetic energy of the hub mass M

$$K_1 = \frac{1}{2}M\left(\dot{x}_b^2 + \dot{y}_b^2\right) + \frac{1}{2}I\dot{\theta}^2 \tag{9.2}$$

The kinetic energy of the end-effector of the manipulator is given by

$$K_2 = \frac{1}{2}m\left(\dot{x}^2 + \dot{y}^2\right) \tag{9.3}$$

The energy stored at the torsional spring is

$$P = \frac{1}{2}k\phi^2 \tag{9.4}$$

Then the total energy of the system is (using (9.1))

$$
\begin{aligned}
E &= K_1 + K_2 + P \\
&= \frac{1}{2}(M + m)\left(\dot{x}^2 + \dot{y}^2\right) + \frac{1}{2}Ml^2\left(\dot{\theta} + \dot{\phi}\right)^2 + \frac{1}{2}I\dot{\theta}^2 \\
&\quad + Ml(\dot{\theta} + \dot{\phi})(\dot{x}\sin(\theta + \phi) - \dot{y}\cos(\theta + \phi)) + \frac{1}{2}k\phi^2 \tag{9.5}
\end{aligned}
$$

Therefore, the Lagrangian function is given by

$$
\begin{aligned}
L &= K_1 + K_2 - P \\
L &= \frac{1}{2}(M + m)\left(\dot{x}^2 + \dot{y}^2\right) + \frac{1}{2}Ml^2\left(\dot{\theta} + \dot{\phi}\right)^2 + \frac{1}{2}I\dot{\theta}^2 \\
&\quad + Ml(\dot{\theta} + \dot{\phi})(\dot{x}\sin(\theta + \phi) - \dot{y}\cos(\theta + \phi)) - \frac{1}{2}k\phi^2 \tag{9.6}
\end{aligned}
$$

The equations of motion are derived using Lagrange's equations

$$\frac{d}{dt}\left(\frac{\partial L}{\partial \dot{q}}(q, \dot{q})\right) - \frac{\partial L}{\partial q}(q, \dot{q}) = \tau \tag{9.7}$$

where $q = (q_1, ...q_n)^T = (x, y, \theta, \phi)^T$ represents the generalized variables, one for each degree of freedom of the system, $\tau = (\tau_1, ..., \tau_n)^T$ denotes forces that are externally applied to the system.

We therefore have

$$\left(\frac{\partial L}{\partial x}\right) = \left(\frac{\partial L}{\partial y}\right) = 0$$

$$\left(\frac{\partial L}{\partial \theta}\right) = Ml(\dot{\theta} + \dot{\phi})\left[\dot{x}\cos(\theta + \phi) + \dot{y}\sin(\theta + \phi)\right]$$

$$\left(\frac{\partial L}{\partial \phi}\right) = \left(\frac{\partial L}{\partial \theta}\right) - k\phi$$

$$\left(\frac{\partial L}{\partial \dot{x}}\right) = (m + M)\dot{x} + Ml\sin(\theta + \phi)(\dot{\theta} + \dot{\phi})$$

$$\left(\frac{\partial L}{\partial \dot{y}}\right) = (m + M)\dot{y} - Ml\cos(\theta + \phi)(\dot{\theta} + \dot{\phi})$$

$$\left(\frac{\partial L}{\partial \dot{\theta}}\right) = I\dot{\theta} + Ml^2(\dot{\theta} + \dot{\phi}) + Ml(\dot{x}\sin(\theta + \phi) - \dot{y}\cos(\theta + \phi))$$

$$\left(\frac{\partial L}{\partial \dot{\phi}}\right) = Ml^2(\dot{\theta} + \dot{\phi}) + Ml(\dot{x}\sin(\theta + \phi) - \dot{y}\cos(\theta + \phi))$$

The virtual work is given by

$$\delta W = F_1\delta(x - l\cos(\theta + \phi)) + F_2\delta(y - l\sin(\theta + \phi)) + T\delta\theta$$

The equations of motion can then be written as follows

$$(M + m)\ddot{x} + Ml(\ddot{\theta} + \ddot{\phi})\sin(\theta + \phi) + Ml(\dot{\theta} + \dot{\phi})^2\cos(\theta + \phi)$$
$$= F_1 \tag{9.8}$$
$$(M + m)\ddot{y} - Ml(\ddot{\theta} + \ddot{\phi})\cos(\theta + \phi) + Ml(\dot{\theta} + \dot{\phi})^2\sin(\theta + \phi)$$
$$= F_2 \tag{9.9}$$
$$Ml(\ddot{x}\sin(\theta + \phi) - \ddot{y}\cos(\theta + \phi)) + (I + Ml^2)\ddot{\theta} + Ml^2\ddot{\phi}$$
$$= l\left(F_1\sin(\theta + \phi) - F_2\cos(\theta + \phi)\right) + T \tag{9.10}$$
$$Ml(\ddot{x}\sin(\theta + \phi) - \ddot{y}\cos(\theta + \phi)) + Ml^2\ddot{\theta} + Ml^2\ddot{\phi} + k\phi$$
$$= l\left(F_1\sin(\theta + \phi) - F_2\cos(\theta + \phi)\right) \tag{9.11}$$

They can also be rewritten in compact form as

$$D(q)\ddot{q} + C(q, \dot{q})\dot{q} + Kq = \tau \tag{9.12}$$

where

$$D(q) = \begin{bmatrix} M + m & 0 & h_s & h_s \\ 0 & M + m & -h_c & -h_c \\ h_s & -h_c & I + Ml^2 & Ml^2 \\ h_s & -h_c & Ml^2 & Ml^2 \end{bmatrix} \tag{9.13}$$

with

$$h_s = Ml\sin(\theta + \phi) \tag{9.14}$$
$$h_c = Ml\cos(\theta + \phi) \tag{9.15}$$

$$C(q,\dot{q}) = \begin{bmatrix} 0 & 0 & Ml\cos(\theta + \phi)(\dot{\theta} + \dot{\phi}) & Ml\cos(\theta + \phi)(\dot{\theta} + \dot{\phi}) \\ 0 & 0 & Ml\sin(\theta + \phi)(\dot{\theta} + \dot{\phi}) & Ml\sin(\theta + \phi)(\dot{\theta} + \dot{\phi}) \\ 0 & 0 & 0 & 0 \\ 0 & 0 & 0 & 0 \end{bmatrix} \tag{9.16}$$

$$K = \begin{bmatrix} 0 & 0 & 0 & 0 \\ 0 & 0 & 0 & 0 \\ 0 & 0 & 0 & 0 \\ 0 & 0 & 0 & k \end{bmatrix} \qquad q = \begin{bmatrix} x \\ y \\ \theta \\ \phi \end{bmatrix} \tag{9.17}$$

and

$$\tau = \begin{bmatrix} F_1 \\ F_2 \\ l(F_1\sin(\theta + \phi) - F_2\cos(\theta + \phi)) + T \\ l(F_1\sin(\theta + \phi) - F_2\cos(\theta + \phi)) \end{bmatrix} \tag{9.18}$$

Note that $D(q)$ is symmetric. Moreover,

$$\det(D(q)) = I(M + m)mMl^2 > 0 \tag{9.19}$$

Therefore $D(q)$ is positive definite for all q. From (9.13) and (9.16), it follows that (using (9.14) and (9.15))

$$\dot{D}(q) - 2C(q,\dot{q}) = (\dot{\theta} + \dot{\phi}) \begin{bmatrix} 0 & 0 & -h_c & -h_c \\ 0 & 0 & -h_s & -h_s \\ h_c & h_s & 0 & 0 \\ h_c & h_s & 0 & 0 \end{bmatrix} \tag{9.20}$$

which is a skew-symmetric matrix. An important property of skew-symmetric matrices, which will be used in establishing the passivity property of the system is

$$z^T(\dot{D}(q) - 2C(q,\dot{q}))z = 0 \qquad \forall z \tag{9.21}$$

9.2.2 Passivity properties of the planar PPR manipulator

The total energy of the robot is given by

$$E = \frac{1}{2}\dot{q}^T D(q)\dot{q} + \frac{1}{2}k\phi^2 \tag{9.22}$$

Therefore, from (9.12)-(9.21), we obtain

$$
\begin{aligned}
\dot{E} &= \dot{q}^T D(q)\ddot{q} + \frac{1}{2}\dot{q}^T \dot{D}(q)\dot{q} + \dot{q}^T Kq \\
&= \dot{q}^T(-C(q,\dot{q})\dot{q} - Kq + \tau) + \frac{1}{2}\dot{q}^T \dot{D}(q)\dot{q} + \dot{q}^T Kq \\
&= \dot{q}^T \tau \\
&= F_1\dot{x} + F_2\dot{y} + l(F_1\sin(\theta + \phi) - F_2\cos(\theta + \phi))(\dot{\theta} + \dot{\phi}) + T\dot{\theta} \\
&= F_1\left[\dot{x} + l\sin(\theta + \phi)(\dot{\theta} + \dot{\phi})\right] + F_2\left[\dot{y} - l\cos(\theta + \phi)(\dot{\theta} + \dot{\phi})\right] + T\dot{\theta} \\
&= \dot{z}^T u \tag{9.23}
\end{aligned}
$$

where u and z are defined as follows: $u = [F_1, \ F_2, \ T]^T$ and $z = [x + l - l\cos(\theta + \phi), \ y - l\sin(\theta + \phi), \ \theta]^T$. Note that $\dot{z} = [\dot{x} + l\sin(\theta + \phi)(\dot{\theta} + \dot{\phi}), \ \dot{y} - l\cos(\theta + \phi)(\dot{\theta} + \dot{\phi}), \ \dot{\theta}]^T$. Integrating both sides of the above equation, we obtain

$$\int_0^t \dot{z}^T u\, dt = E(t) - E(0) \tag{9.24}$$

Therefore, the system having u as input and $\dot{z}$ as output is passive.

Our control objective is to move the system from any given initial configuration $(x^0, y^0, \theta^0, \phi^0)$ to the origin $(0, 0, 0, 0)$.

9.3 Energy-based stabilizing control law

The passivity property of the system suggests the use of the total energy E in (9.22) and the vector z, in the controller design. We wish to bring E to zero. Note that E is a positive definite function. We propose the following Lyapunov function candidate

$$
\begin{aligned}
V(q, \dot{q}) &= k_E E(q, \dot{q}) + \frac{1}{2}z^T Qz \\
&= k_E E(q, \dot{q}) + \frac{k_x}{2}(x + l - l\cos(\theta + \phi))^2 \\
&\quad + \frac{k_y}{2}(y - l\sin(\theta + \phi))^2 + \frac{k_\theta}{2}\theta^2 \tag{9.25}
\end{aligned}
$$

where $Q = \begin{bmatrix} k_x & 0 & 0 \\ 0 & k_y & 0 \\ 0 & 0 & k_\theta \end{bmatrix}$ and where k_E, k_x, k_y, k_θ are strictly positive
constants. Note that $V(q, \dot{q})$ is a positive definite function. Taking the
time derivative of V and using (9.23), we obtain

$$
\begin{aligned}
\dot{V} &= k_E \dot{E} + k_x (x + l - l \cos(\theta + \phi))[\dot{x} + l \sin(\theta + \phi)(\dot{\theta} + \dot{\phi})] \\
&\quad + k_y (y - l \sin(\theta + \phi))[\dot{y} - l \cos(\theta + \phi)(\dot{\theta} + \dot{\phi})] + k_\theta \theta \dot{\theta} \\
&= [\dot{x} + l \sin(\theta + \phi)(\dot{\theta} + \dot{\phi})](k_E F_1 + k_x(x + l - l \cos(\theta + \phi))) \\
&\quad + [\dot{y} - l \cos(\theta + \phi)(\dot{\theta} + \dot{\phi})](k_E F_2 + k_y(y - l \sin(\theta + \phi))) \\
&\quad + \dot{\theta}(k_E T + k_\theta \theta)
\end{aligned}
\tag{9.26}
$$

We propose the control inputs as follows

$$
T = \frac{1}{k_E}[-k_\theta \theta - \dot{\theta}]
\tag{9.27}
$$

$$
\begin{aligned}
F_1 &= \frac{1}{k_E}[k_x(-x - l + l \cos(\theta + \phi)) - \dot{x} \\
&\quad - l \sin(\theta + \phi)(\dot{\theta} + \dot{\phi})]
\end{aligned}
\tag{9.28}
$$

$$
F_2 = \frac{1}{k_E}[k_y(-y + l \sin(\theta + \phi)) - \dot{y} + l \cos(\theta + \phi)(\dot{\theta} + \dot{\phi})]
\tag{9.29}
$$

Introducing the above in (9.26), one has

$$
\begin{aligned}
\dot{V} &= -(\dot{x} + l \sin(\theta + \phi)(\dot{\theta} + \dot{\phi}))^2 \\
&\quad - (\dot{y} - l \cos(\theta + \phi)(\dot{\theta} + \dot{\phi}))^2 - \dot{\theta}^2 \\
&= -\dot{z}^T \dot{z}
\end{aligned}
\tag{9.30}
$$
$$
\tag{9.31}
$$

9.3.1   Equivalent closed-loop interconnection

This section gives the equivalent closed-loop interconnection of the system and refers, among others, to the work of [14, 56, 57]. Note that this interpretation is a possible way to study stability properties of the system. Indeed, we can interpret the dynamics of the system as the negative interconnection of two passive blocks and then using the passivity theorem, we can conclude on stability.

Looking at the closed-loop system (9.12) using the control inputs (9.27), (9.28) and (9.29) we can interpret these dynamics as the interconnection of two subsystems with respect to inputs vector u_1, u_2 and outputs vector y_1 and y_2, with $y_1 = u_2$ and $y_2 = -u_1$, and

$$\begin{cases} u_1 = u \\ y_1 = \dot{z} \end{cases} \tag{9.32}$$

where $u = [F_1 \ F_2 \ T]^T$ with (9.27), (9.28) and (9.29) and $\dot{z} = [\dot{x} + l\sin(\theta + \phi)(\dot{\theta} + \dot{\phi}) \quad \dot{y} - l\cos(\theta + \phi)(\dot{\theta} + \dot{\phi}) \quad \dot{\theta}]^T$. This is motivated by the fact that the planar PPR manipulator dynamics in (9.12), with the control inputs in (9.27), (9.28) and (9.29) defines a passive operator between u_1 and y_1 (see (9.24)), with state vector $\begin{pmatrix} q \\ \dot{q} \end{pmatrix}$ and dynamics

$$D(q)\ddot{q} + C(q, \dot{q})\dot{q} + Kq = \tau \tag{9.33}$$

Let us write the second subsystem in state space form as

$$\begin{cases} \dot{z} = u_2 \\ y_2 = H(s)u_2 \end{cases} \tag{9.34}$$

where the transfer matrix $H(s)$ is given by

$$H(s) = \begin{bmatrix} \frac{k_x + s}{k_E s} & 0 & 0 \\ 0 & \frac{k_y + s}{k_E s} & 0 \\ 0 & 0 & \frac{k_\theta + s}{k_E s} \end{bmatrix} \tag{9.35}$$

Thus, $H(s)$ is positive real (PR) (see Definition 2.3, in [56]). Applying the passivity theorem, it follows that $\dot{z} \in L^2(\mathbb{R}^+)$. Notice that this is a consequence of the fact that $H(s)$ defines an input strictly passive operator, see Theorem 2.3 2), in [56]. The storage functions of each subsystem are equal to $k_E E$ for the first subsystem and $\frac{1}{2}z^T Q z$ for the second subsystem. The sum of both storage functions yields the desired Lyapunov function defined in (9.25). The interconnection is depicted in Figure 9.2.

What is important is that we can systematically associate with these dissipative subsystems some Lyapunov functions that are systematically deduced from their passivity property. This is a fundamental property of dissipative systems that one can use to calculate Lyapunov functions for them.

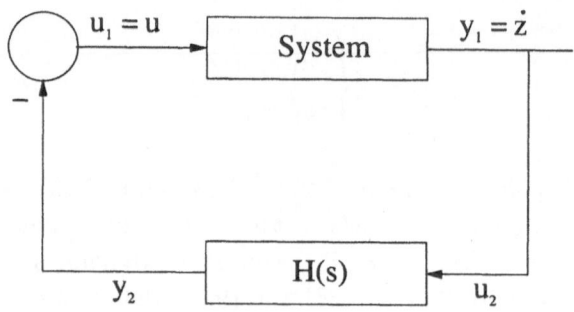

Figure 9.2: The equivalent representation as the negative feedback interconnection of a passive system and a SPR transfer function

Note also that all the fundamental theory on dissipative systems is extensively related to the well-known Kalman-Yakubovich-Popov lemma, which is one of the key results in control theory.

The main result is stated in the following theorem.

Theorem 9.1 *Consider the planar PPR robot (9.12). Taking the Lyapunov function candidate (9.25) with strictly positive constants k_E, k_x, k_y and k_θ, then the solution of the closed-loop system (x, y, θ, ϕ) with the control inputs (9.27), (9.28) and (9.29) converges to the origin $(0, 0, 0, 0)$.* ∎

The proof will be developed in the following section.

9.4   Convergence and stability analysis

We will mainly use LaSalle's invariance theorem to prove the above theorem. In order to apply LaSalle's theorem, we are required to define a compact (closed and bounded) set Ω with the property that every solution of system (9.12) that starts in Ω remains in Ω for all future time. Recall that the Lyapunov function candidate V is positive definite. Since $V(q, \dot{q})$ in (9.25) is a decreasing function, (see (9.31)), then q and $\dot{q}$ are bounded. Therefore, the solution of the closed-loop system remains inside a compact set Ω that is defined by the initial state values. Let Γ be the set of all points in Ω such that $\dot{V}(q, \dot{q}) = 0$. Let M be the largest invariant set in Γ. LaSalle's theorem ensures that every solution starting in Ω approaches M as $t \to \infty$. Let us now compute the largest invariant set M in Γ.

In the set Γ (see (9.31)), $\dot{V} = 0$ and thus

$$\dot{\theta} = 0 \tag{9.36}$$
$$\ddot{x} + l\sin(\theta + \phi)(\dot{\theta} + \dot{\phi}) = 0 \tag{9.37}$$
$$\ddot{y} - l\cos(\theta + \phi)(\dot{\theta} + \dot{\phi}) = 0 \tag{9.38}$$

$\dot{\theta} = 0$ implies that θ is constant. Subtracting equation (9.11) from equation (9.10) yields

$$I\ddot{\theta} - k\phi = T \tag{9.39}$$

Since $\dot{\theta} = 0$, then $\ddot{\theta} = 0$ and using (9.39), we have $T = -k\phi$. Moreover, since T has been chosen such that $T = \frac{1}{k_E}[-k_\theta\theta - \dot{\theta}]$ (see (9.27)) and since $\dot{\theta} = 0$, then $k_\theta\theta = k_E k\phi$. Therefore, ϕ is constant. So, $\dot{\phi} = 0$ and $\ddot{\phi} = 0$.

From (9.37), (9.38), $\dot{\theta} = 0$ and $\dot{\phi} = 0$, it then follows that $\ddot{x} = 0$ and $\ddot{y} = 0$. Therefore, x, y, θ and ϕ are constants.

In view of (9.8), $F_1 = 0$ and in view of (9.9), $F_2 = 0$. From (9.11), it follows that $\phi = 0$. Using $k_\theta\theta = k_E k\phi$, it follows that $\theta = 0$. Then, from (9.28), (9.29), (9.37) and (9.38), we conclude that $x = 0$ and $y = 0$ also.

Finally, the largest invariant set M is given by the origin $(0, 0, 0, 0)$. All the solutions of the closed-loop system converge to the origin. This ends the proof of Theorem 9.1.

9.5 Simulation results

In order to observe the performance of the proposed control law based on passivity, we performed simulations on MATLAB, using SIMULINK.

We considered the system taking the parameters as follows: $M = 2$, $m = 1$, $l = 3$, $I = 1$ and $k = 1$. Figures 9.3 and 9.4 show the results for $k_E = 1$, $k_x = 1$, $k_y = 1$, $k_\theta = 1$ and for an initial position

$$
\begin{array}{cccc}
x = 5 & y = 1 & \theta = 1 & \phi = 0 \\
\dot{x} = 0, & \dot{y} = 0 & \dot{\theta} = 0 & \dot{\phi} = 0
\end{array}
$$

Simulations showed that our control law brings the state of the system to the origin. The control inputs go to zero and the Lyapunov function V is always decreasing and converges to zero.

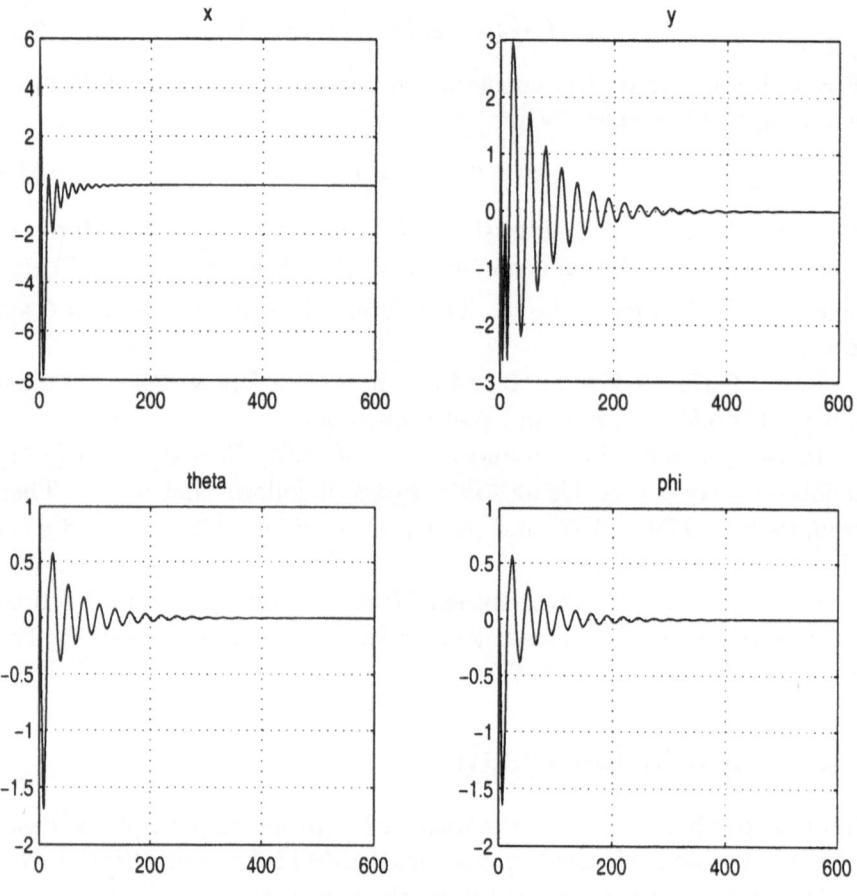

Figure 9.3: States of the system

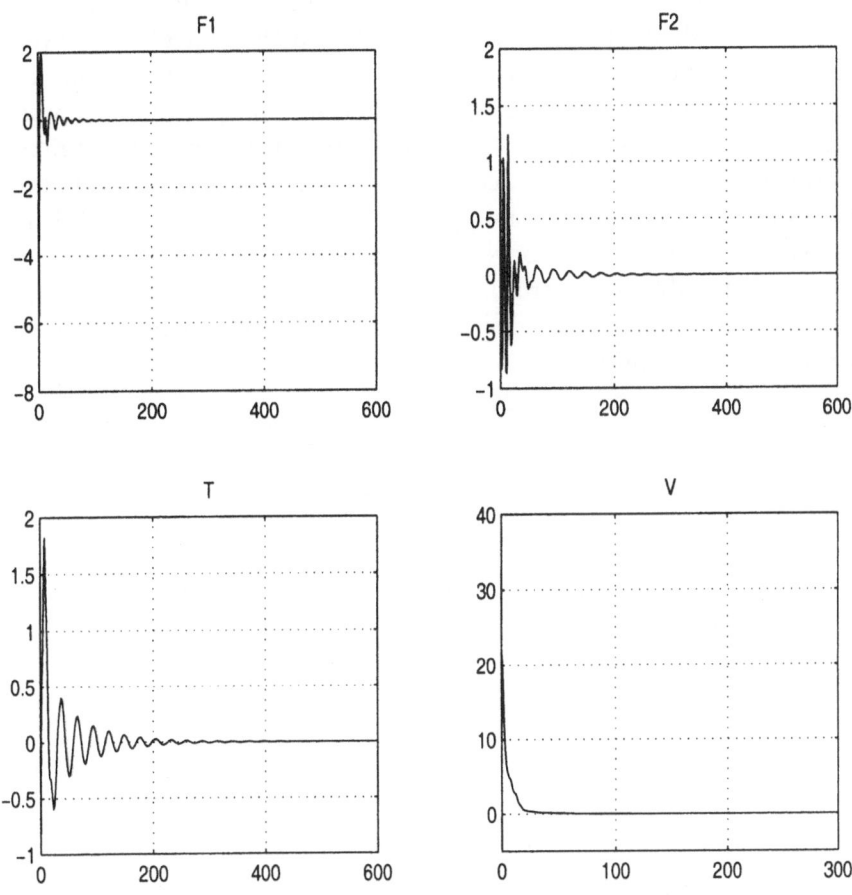

Figure 9.4: Control inputs and Lyapunov function V

9.6   Conclusions

The control strategy presented in this chapter brings the state of the planar PPR manipulator to the origin. The control strategy is based on an energy approach and the passivity properties of the system. A Lyapunov function is obtained using the total energy of the system. The analysis is carried out using LaSalle's invariance theorem. In contrast to the works of [8] and [88] in which they consider the control of the system without excitation of its elastic mode, we have proposed a stabilizing control law using the elastic energy. Indeed, our main interest was to extend the previous works [18, 21, 24, 59] to this underactuated manipulator, which possesses four degrees of freedom and three control inputs. We can notice that the passivity properties are essential in establishing the control law. In the above previous works, the systems were controlled with only one control input. This particular system proves that our approach can be extended to some underactuated systems having several control inputs.

Chapter 10

The ball and beam acting on the ball

10.1 Introduction

Another interesting example of underactuated systems is the ball and beam system. In this chapter*, we propose a control law for the ball and beam system acting on the ball instead of the beam. Such a mechanical system is outlined in Figure 10.1 and described in Section 10.2. The system in Figure 10.1 is motivated by the control of small rotational oscillations of platforms and vehicle suspensions. For simplicity, we assume that the center of the beam is connected to the pivot using a rotational spring and we neglect friction.

We follow the Lagrangian approach to obtain the dynamical model of the system and present some important mechanical properties such as passivity (see [117] and [46]). We propose a feedback control scheme, which allows us to asymptotically stabilize the system from any initial condition close enough to the equilibrium point.

In Section 10.2, we describe the kinematics of the ball and beam system when a force is acting on the ball. We obtain the dynamical model using an Euler-Lagrange formulation and show the passivity properties of the system. In Section 10.3, we derive the input control law and prove the stability of the closed-loop system. In Section 10.4, we present some

---

*The authors of this chapter are Carlos Aguilar and Rogelio Lozano. Carlos Aguilar is with the Laboratory of Measurement and Control, CIC, IPN, Col. San Pedro Zacatenco, AP. 75476, 07700 Mexico D.F. and Rogelio Lozano is with the Laboratory Heudiasyc, UTC UMR CNRS 6599, Centre de Recherche de Royallieu, BP 20529, 60205 Compiègne Cedex, France.

numerical simulations. Section 10.5 is devoted to conclusions and some suggestions for future research.

10.2 Dynamical model

Let us consider the dynamical system described in Figure 10.1. We are dealing basically with the ball and beam system but the control input is now the force acting on the ball. A rotational spring has been added, which is represented in the figure as two springs to remind the reader about the original motivation of the system, which is a vehicle's suspension scheme.

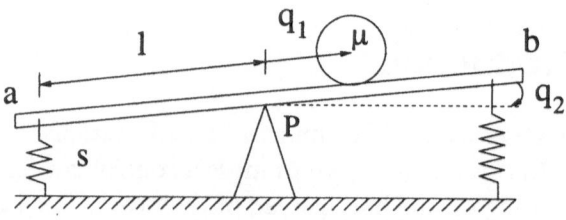

Figure 10.1: The ball and beam system

To describe the motion, we choose the origin of the reference system at point P and set the x axis along the horizontal direction and the y axis along the vertical direction. As a set of generalized coordinates, we use q_2 as the angle between the beam and the x axis, and q_1 as the displacement of the ball from the origin P and measured along the beam.

The position vector of the center of mass of the ball and its velocity are

$$\vec{R}_M = q_1(\cos q_2, \sin q_2) \tag{10.1}$$
$$\vec{V}_M = (-q_1\dot{q}_2 \sin q_2 + \dot{q}_1 \cos q_2, q_1\dot{q}_2 \cos q_2 + \dot{q}_1 \sin q_2) \tag{10.2}$$

Let μ be the mass of the ball, I_μ the moment of inertia of the ball around its center of mass and $I_B = \frac{2}{3}ml^2$ the moment of inertia of the beam. Using equation (10.2), the total kinetic energy of the system is given by

$$K(q, \dot{q}) = \frac{1}{2}I_T\dot{q}_2^2 + \frac{1}{2}\mu\left(\dot{q}_1^2 + q_1^2\dot{q}_2^2\right) \tag{10.3}$$

where $I_T = I_B + I_\mu$. The potential energy of the ball is

$$V_\mu(q) = \mu g q_1 \sin q_2$$

and the potential energy of the deformed springs is

$$V_S(q) = k l^2 \sin^2 q_2$$

where k is the elastic constant of the springs. Therefore, the total potential energy is given by

$$V(q) = V_S(q) + V_\mu(q) = k l^2 \sin^2 q_2 + \mu g q_1 \sin q_2 \qquad (10.4)$$

Equations (10.3) and (10.4) allow us to write down the Lagrangian function

$$L(q, \dot{q}) = K(q, \dot{q}) - V(q) \qquad (10.5)$$

where $q^T = [q_1 \ q_2]$.

Following the Euler-Lagrange procedure, we get two second order differential equations as follows

$$\frac{d}{dt}\left(\frac{\partial L}{\partial \dot{q}_1}\right) - \frac{\partial L}{\partial q_1} = \mu \ddot{q}_1 - \mu q_1 \dot{q}_2^2 + \mu g \sin q_2 \qquad (10.6)$$

$$\frac{d}{dt}\left(\frac{\partial L}{\partial \dot{q}_2}\right) - \frac{\partial L}{\partial q_2} = (\mu q_1^2 + I_T)\ddot{q}_2 + 2\mu q_1 \dot{q}_1 \dot{q}_2$$

$$+ 2k l^2 \cos q_2 \sin q_2 + \mu g q_1 \cos q_2 \qquad (10.7)$$

In the general case, these equations can be written as

$$M(q)\ddot{q} + C(q, \dot{q})\dot{q} + G(q) + F_s(q) = \tau \qquad (10.8)$$

where $M(q)$ is the inertia matrix

$$M(q) = \begin{bmatrix} \mu & 0 \\ 0 & \mu q_1^2 + I_T \end{bmatrix} \qquad (10.9)$$

$C(q, \dot{q})$ is the Coriolis matrix

$$C(q, \dot{q}) = \begin{bmatrix} 0 & -\mu q_1 \dot{q}_2 \\ \mu q_1 \dot{q}_2 & \mu q_1 \dot{q}_1 \end{bmatrix} \qquad (10.10)$$

$G(q)$ and $F_S(q)$ are given by

$$G(q) = \begin{bmatrix} \mu g \sin q_2 \\ \mu g q_1 \cos q_2 \end{bmatrix} \qquad F_s(q) = \begin{bmatrix} 0 \\ 2k l^2 \cos q_2 \sin q_2 \end{bmatrix} \qquad (10.11)$$

Note that $G(q)$ and $F_s(q)$ are due to gravity and to the spring respectively.

The generalized force is given by

$$\tau = \begin{bmatrix} f \\ w \end{bmatrix} \tag{10.12}$$

where f is the force acting on the ball along the beam and w is an external perturbation.

It can be easily seen that equations (10.8) to (10.12) define an underactuated system, because the system has only one input f and two degrees of freedom q_1 and q_2.

It is worth mentioning that if the control input is zero, i.e. $f = 0$, the disturbance w can render the system unstable.

10.2.1 Mechanical properties

The mechanical system (10.8) has several fundamental properties, which can be used to facilitate the design of a control system.

- $\underline{P.1}$: $M(q)$ is a positive definite matrix.

- $\underline{P.2}$: $N = \dot{M}(q) - 2C(q, \dot{q})$ is a skew-symmetric matrix

$$N = \begin{bmatrix} 0 & 2\mu q_1 \dot{q}_2 \\ -2\mu q_1 \dot{q}_2 & 0 \end{bmatrix} \tag{10.13}$$

Therefore

$$q^T N q = 0 \quad \forall \quad q \in \mathbb{R}^2 \tag{10.14}$$

- $\underline{P.3}$: Vectors $G(q)$ and $F_S(q)$ satisfy the following relation (see (10.4))

$$\frac{\partial V(q)}{\partial q} = G(q) + F_s(q)$$

Remark 10.1 *Properties P.1 and P.3 allow us to propose a Lyapunov function that is related to the total energy. Property P.2 is a statement about the Coriolis forces, which will be useful when proving negativeness of the derivative of the Lyapunov function.* ∎

10.3 The control law

Property $P.2$ suggests the use of the total energy to design an input control law, which stabilizes the system at the equilibrium point $(q^T, \dot{q}^T) = (0,0,0,0)$, starting from any initial condition $(q_{10}, q_{20}, \dot{q}_{10}, \dot{q}_{20})$. To solve this problem, we propose a simple passive feedback scheme, which corresponds to a PD controller defined by

$$f = -k_p q_1 - k_d \dot{q}_1 \qquad (10.15)$$

where $k_p > 0$ and $k_d > 0$.

10.3.1 Stability analysis

We obtain the closed-loop equations by introducing the PD control (10.15) into (10.8) for the case when $w = 0$

$$\ddot{q}_1 = -g \sin q_2 + q_1 \dot{q}_2^2 - \frac{(k_d \dot{q}_1 + k_p q_1)}{\mu} \qquad (10.16)$$

$$\ddot{q}_2 = \frac{-2\mu q_1 \dot{q}_1 \dot{q}_2 - 2kl^2 \cos q_2 \sin q_2 - \mu g q_1 \cos q_2}{\mu q_1^2 + I_T} \qquad (10.17)$$

Before stating the main result of this paper, we present a useful proposition.

Proposition 10.1 *Let us consider the following auxiliary function*

$$V_P(q) = V(q) + \frac{k_p q_1^2}{2} \qquad (10.18)$$

where $V(q)$ is given in (10.4). If $k_p > \frac{\mu^2 g^2}{2kl^2}$, then $V_P(q)$ is positive definite for all $q \in \mathbb{R}^2$. ∎

Proof 10.1 *Note that*

$$V_P(q) = kl^2 \sin^2 q_2 + \mu g q_1 \sin q_2 + \frac{k_p q_1^2}{2}$$

By applying the following inequality: $-ab \leq \frac{a^2}{2}\gamma + \frac{b^2}{2\gamma} \ \forall a, b \in \mathbb{R}, \ \gamma > 0$ into the above equation, we then have

$$V_P(q) \geq (kl^2 - \frac{\gamma \mu g}{2}) \sin^2 q_2 + \frac{q_1^2}{2}(k_p - \frac{\mu g}{\gamma}) \qquad (10.19)$$

It follows that $V_p(q)$ will be positive definite if $\gamma < \frac{2kl^2}{\mu g}$ and if k_p is chosen such that $k_p > \frac{\mu g}{\gamma}$. Therefore, $V_p(q)$ is positive definite if

$$k_p > \frac{\mu^2 g^2}{2kl^2} \tag{10.20}$$

∎

Define the following Lyapunov function candidate

$$V_T(q, \dot{q}) = \frac{1}{2} \dot{q}^T M(q) \dot{q} + V_P(q) \tag{10.21}$$

where V_P is given in (10.18). If k_p verifies (10.20), then $V_T(q, \dot{q})$ is positive definite.

The time derivative of V_T is

$$\dot{V}_T = \dot{q}^T M(q) \ddot{q} + \frac{1}{2} \dot{q}^T \frac{dM(q)}{dt} \dot{q} + \dot{q}^T \frac{\partial V(q)}{\partial q} + k_p q_1 \dot{q}_1$$

From (10.8) and properties $P.1$ to $P.3$, we get

$$
\begin{aligned}
\dot{V}_T &= \dot{q}^T \left(-C(q, \dot{q})\dot{q} - G(q) - F_s(q) + \tau \right) \\
&\quad + \frac{1}{2} \dot{q}^T \frac{dM(q)}{dt} \dot{q} + \dot{q}^T \frac{\partial V(q)}{\partial q} + k_p q_1 \dot{q}_1 \\
&= \dot{q}_1 f + k_p q_1 \dot{q}_1 \tag{10.22}
\end{aligned}
$$

Introducing f in (10.15) gives

$$\dot{V}_T = -k_d \dot{q}_1^2 \tag{10.23}$$

Now we proceed to apply LaSalle's theorem. Since $V_T(q, \dot{q})$ is positive definite in $\mathbb{R}^4$, while $\dot{V}_T(q, \dot{q})$ is only negative semi-definite, stability in the sense of Lyapunov is guaranteed, i.e. q and $\dot{q}$ are bounded. The asymptotic stability of the equilibrium point ($q_1 = 0, q_2 = 0, \dot{q}_1 = 0, \dot{q}_2 = 0$) follows from LaSalle's invariance theorem. Define the set (see [46]))

$$S = \left\{ \dot{V}_T(q, \dot{q}) = 0 \right\} = \{ \dot{q}_1 = 0 \} \tag{10.24}$$

and consider any trajectory $(q, \dot{q})$ inside S, which is an invariant set, since $\dot{q}_1 = 0$ then $q_1 = \bar{q}_1$ where $\bar{q}_1$ is a constant. From (10.16), it follows that

$$0 = -\mu g \sin q_2 + \mu \bar{q}_1 \dot{q}_2^2 - k_p \bar{q}_1 \tag{10.25}$$

Differentiating the above expression, we get

$$0 = \dot{q}_2(-\mu g \cos q_2 + 2\mu \bar{q}_1 \ddot{q}_2) \tag{10.26}$$

We then have two possible cases:

- Case a: $\dot{q}_2 = 0$.

  From (10.25), it follows that q_2 is constant, i.e. $q_2 = \bar{q}_2$ and such that

  $$\sin \bar{q}_2 + \frac{k_p}{\mu g} \bar{q}_1 = 0$$

  From (10.17), it follows that

  $$\cos \bar{q}_2 (\sin \bar{q}_2 + \frac{\mu g}{2kl^2} \bar{q}_1) = 0$$

  Combining the two above equations, we have

  $$\cos \bar{q}_2 (\frac{\mu^2 g^2}{2kl^2} - k_p) \bar{q}_1 = 0$$

  In view of (10.20), we either have $\bar{q}_1 = 0$ or $\cos \bar{q}_2 = 0$. If $\bar{q}_1 = 0$, then from (10.25) $\bar{q}_2 = 0$. In order to avoid having $\bar{q}_2 = \pm \frac{\pi}{2}$ as possible convergence points, it is sufficient that the initial conditions belong to a neighbourhood of the origin such that (see (10.19))

  $$V(0) < kl^2 - \frac{\gamma \mu g}{2} \tag{10.27}$$

  for some $\gamma > 0$ such that the right hand side of the above is positive and $k_p > \frac{\mu g}{\gamma}$.

- Case b: $\dot{q}_2 \neq 0$, then

  $$\ddot{q}_2 = \frac{g}{2\bar{q}_1} \cos q_2$$

  Combining the above with (10.17), we get

  $$\cos q_2 \left(\frac{g}{2\bar{q}_1} (\mu \dot{q}_1^2 + I_T) + 2kl^2 \sin q_2 + \mu g \bar{q}_1 \right) = 0 \tag{10.28}$$

  The case $\cos q_2 = 0$ has been avoided by imposing (10.27). It turns out that $q_2 = \bar{q}_2$ is constant. From (10.25), it follows that

  $$\sin \bar{q}_2 = -\frac{k_p}{\mu g} \bar{q}_1 \tag{10.29}$$

From (10.28) and (10.29), we get

$$\frac{2kl^2}{\mu g \bar{q}_1} \left(\frac{\mu^2 g^2}{4kl^2}(\bar{q}_1^2 + \frac{I_T}{\mu}) + \bar{q}_1^2(\frac{\mu^2 g^2}{2kl^2} - k_p) \right) = 0 \qquad (10.30)$$

If k_p satisfies the inequality

$$\frac{1}{2}\frac{\mu^2 g^2}{kl^2} < k_p \le \frac{3}{4}\frac{\mu^2 g^2}{kl^2} \qquad (10.31)$$

then (10.30) leads to a contradiction ruling out the possibility of having case b).

For simplicity, let us choose $\gamma = \frac{8}{5}\frac{kl^2}{\mu g}$. Inequality (10.27) becomes

$$V(0) < \frac{kl^2}{5}. \qquad (10.32)$$

From (10.19) k_p has to satisfy $k_p > \frac{\mu g}{\gamma}$, which becomes

$$k_p > \frac{\mu g}{\gamma} = \frac{5}{8}\frac{\mu^2 g^2}{kl^2} \qquad (10.33)$$

Therefore we can choose

$$k_p = \frac{3}{4}\frac{\mu^2 g^2}{kl^2} \qquad (10.34)$$

which satisfies (10.31) and (10.33).

We can summarize our results in the following lemma.

Lemma 10.1 *Let us consider the ball and beam system as described by (10.8) in closed-loop with the controller (10.15) with k_p as in (10.34). Then, the origin of the system is locally asymptotically stable and the domain of attraction is the region defined by (10.32) where the Lyapunov function is given in (10.21).* ∎

10.4   Simulation results

In order to observe the performance of the feedback control law proposed in equation (10.15), we performed some simulations using SIMNON. We considered:

$$\mu = 1, \quad I_T = 10, \quad l = 1, \quad k_p = 5, \quad k_d = 3.$$

Figures 10.2, 10.3 and 10.4 show the behavior of q_1, q_2 and the control action f, respectively, starting from $q_1(0) = 0.6$, $q_2(0) = 0.15$, $\dot{q}_1(0) = -0.1$ and $\dot{q}_2(0) = 0.1$. As we can see, this system exhibits good behavior. The stabilization occurs around 50 seconds with a control effort that has an upper bound of $|f| < 2N$.

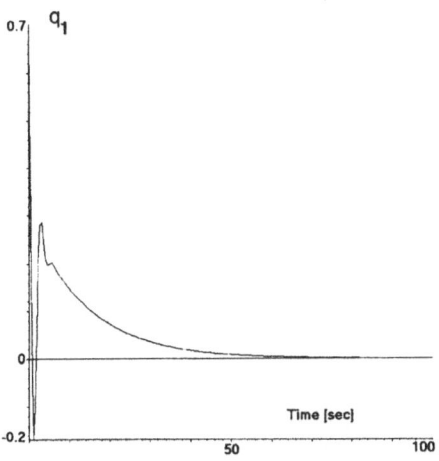

Figure 10.2: Displacement q_1

In order to evaluate the robustness of the closed-loop performance against an external disturbance w, we applied a disturbance w (see (10.12)) of amplitude of $1N$, during 0.5 seconds, once the system had reached the equilibrium point. A plot of the corresponding response is shown in Figures 10.5, 10.6 and 10.7. We note that the amplitudes of q_1, q_2 and the control action f are smaller than those shown in the previous figures.

10.5 Conclusions

In this chapter, we have presented a control strategy for the ball and beam system by acting on the ball, which exploits the natural passivity properties of this class of underactuated systems. The objective of the control law was to bring the state to zero, from any initial condition $(q_{10}, q_{20}, \dot{q}_{10}, \dot{q}_{20})$ close enough to the equilibrium point. We considered the movement of the system subject to small vibrations around the equilibrium point.

The stability of the closed-loop system was shown using the second

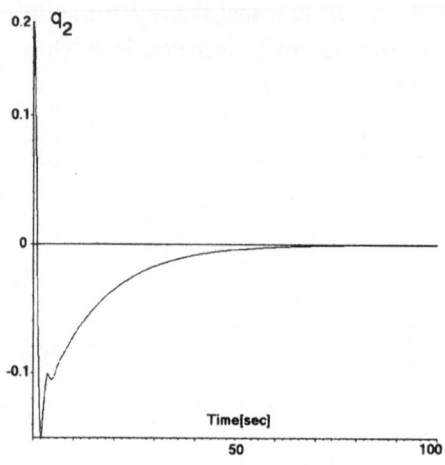

Figure 10.3: Angle q_2

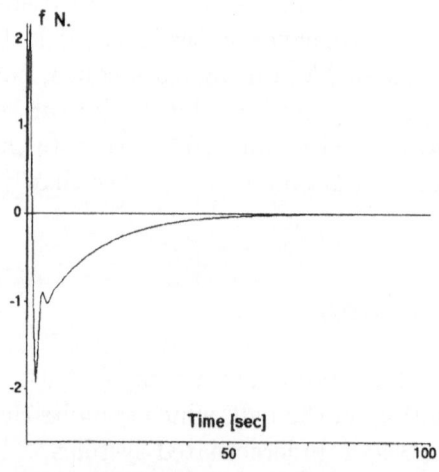

Figure 10.4: Input f

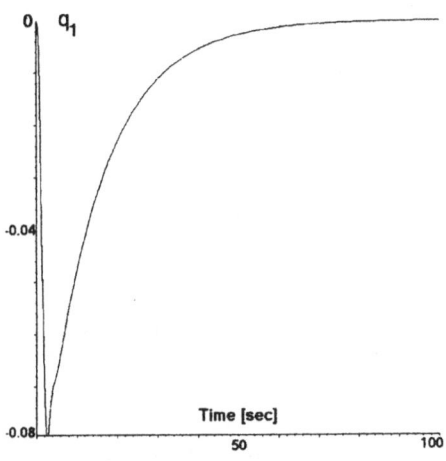

Figure 10.5: Displacement q_1

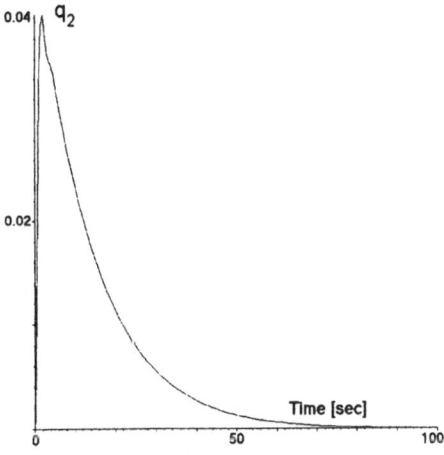

Figure 10.6: Angle q_2

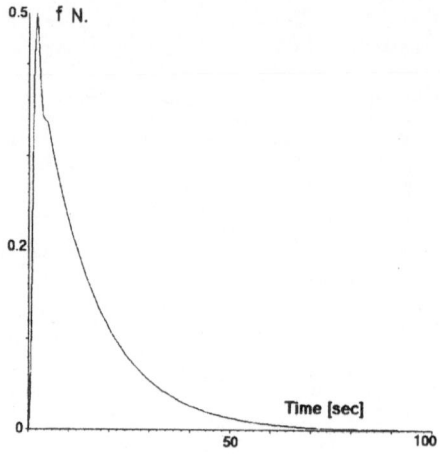

Figure 10.7: Input f

Lyapunov method. We proposed a Lyapunov function V_T based on the total energy of the system. We used the mechanical properties of Euler-Lagrange systems. The convergence analysis was made using LaSalle's theorem, from which we guarantee that the closed-loop system is asymptotically stable.

Chapter 11

The hovercraft model

11.1 Introduction

Nowadays, control problems of underactuated vehicles motivate the development of new non-linear control design methodologies. Such systems are vehicles with fewer independent control inputs than degrees of freedom to be controlled.

In order to capture the essential non-linear behavior of an underactuated ship, we have simplified its model as found in [25]. Neglecting the damping, we have considered that the shape of the ship is symmetric with respect to three axes, mainly a circle and that the two propellers are situated at the center of mass. Therefore, after these simplifications, we obtain the model of a hovercraft that has two propellers to move the vehicle forwards (and backwards) and to make it turn. The main difference with respect to a two-wheel mobile robot is that a hovercraft can move freely sideways, even though this degree of freedom is not actuated. The hovercraft model presented here will be used to design a control strategy and the purpose is to promote the development of new control design methods, such as the studies of other highly non-linear mechanical systems like the ball and beam and the inverted pendulum have done.

A picture of a model kit representing a real hovercraft (the "LCAC-1 Navy Assault Hovercraft") is shown in Figure 11.1. The Landing Craft Air Cushion (LCAC-1) is an assault vehicle designed to transport U.S. Marine fighting forces from naval ships off shore to inland combat positions. The model kit is a reproduction of the 200 ton craft.

We will first consider the problem of regulating the surge, the sway and the angular velocities to zero. We will also propose strategies for

Figure 11.1: LCAC-1 Navy Assault Hovercraft

positioning the hovercraft at the origin.

Various control algorithms for controlling underactuated vessels have appeared in the literature. Leonard [51] was the first to control a dynamic autonomous underwater vessel model (AUV model) with force and torque control inputs. It was shown how open-loop periodic time-varying control can be used to control underwater vehicles. Pettersen and Egeland [82] developed a stability result involving continuous time-varying feedback laws that exponentially stabilize both the position and orientation of a surface vessel having only two control inputs. This result was extended to include integral action in [83]. Other approaches also exist in the literature like the one in Fossen et al. [26], which considered a non-linear ship model including the hydro-dynamic effects due to time-varying speed and wave frequency. This involved a non-symmetrical inertia matrix and non-positive damping at high-speed. The authors used a backstepping technique for tracking control design. Bullo and Leonard [15] developed high-level motion procedures that solved point-to-point reconfiguration, local exponential stabilization and static interpolation problems for underactuated vehicles. Strand et al. [112] proposed a stabilizing controller for moored and free-floating (but not underactuated) ships constructed by backstepping. They proposed a locally asymptotically convergent algorithm based on H_∞-optimal control. They also presented a global result using inverse optimality for the non-linear system. Pettersen and Nijmeijer [85] proposed a time-varying feedback control law that provides global practical stabilization and tracking, using a combined integrator back-stepping and averaging approach. In [84], they proposed a tracking control law that steers both position and course angle of the surface vessel, providing semi-global exponential stabilization of the desired trajectory. Berge et al. [9] developed a tracking controller for the underactuated ship using partial feedback linearization. The control law makes the position and velocities converge exponentially to the reference trajectory, while the course angle of the ship is not controlled.

One of the difficulties encountered in the stabilization of underactuated vehicles is that classical non-linear techniques in non-linear control theory like feedback linearization are not applicable. Therefore, new design methodologies should be explored.

In the present chapter, we propose two different control strategies. The first controller globally and asymptotically stabilize the surge, sway and angular velocities with a differentiable controller. In this case, we consider the surge force and the angular torque as inputs. In the second controller, we globally and asymptotically stabilizes the position and the sway velocities at the origin using the surge and the angular velocities as inputs. The proposed controller is discontinuous. In both cases, the analysis is based on a Lyapunov approach. This chapter refers to the work [22]. The chapter is organized as follows. In Section 11.2, the model of the simplified ship is recalled. Section 11.3 presents the control algorithm to stop the hovercraft. Section 11.4 is devoted to the control strategy for positioning of the hovercraft. Section 11.5 gives simulation results.

11.2   The hovercraft model

In this section, the mathematical model of the hovercraft system as shown in Figure 11.2 is obtained using both Newton's second law and the Euler-Lagrange formulation.

11.2.1   System model using Newton's second law

We consider the class of underactuated vehicles described by the following general model (see [25])

$$M\dot{\nu} + C(\nu)\nu + D(\nu)\nu + g(\eta) = \begin{bmatrix} \tau \\ 0 \end{bmatrix} \qquad (11.1)$$

$$\dot{\eta} = J(\eta)\nu \qquad (11.2)$$

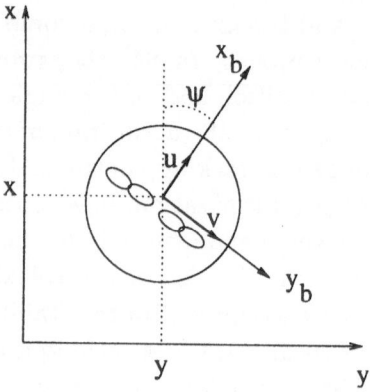

Figure 11.2: The hovercraft

where $\eta \in \mathbb{R}^n$, $\nu \in \mathbb{R}^n$, $\tau \in \mathbb{R}^m$, $m < n$. The matrices M and J are non-singular and $\dot{M} = 0$. This class of systems includes underactuated surface vessels, underwater vehicles, aeroplanes and spacecraft. The vector ν denotes the linear and angular velocities decomposed in the body-fixed frame, η denotes the position and orientation decomposed in the earth-fixed frame, and τ denotes the control forces and torques decomposed in the body-fixed frame. M is the inertia matrix. $C(\nu)$ is the Coriolis and centripetal matrix. $D(\nu)$ is the damping matrix and $g(\eta)$ is the vector of gravitational and possibly buoyant forces and torques. Equations (11.1) and (11.2) represent the dynamics and the kinematics respectively.

Using the previous model (11.1)-(11.2), a surface vessel having two independent main propellers is described by the following model (see [81])

$$
\begin{bmatrix} m_{11} & 0 & 0 \\ 0 & m_{22} & m_{23} \\ 0 & m_{23} & m_{33} \end{bmatrix} \begin{bmatrix} \dot{u} \\ \dot{v} \\ \dot{r} \end{bmatrix} + \begin{bmatrix} 0 & 0 & -f(v,r) \\ 0 & 0 & m_{11}u \\ f(v,r) & m_{11}u & 0 \end{bmatrix} \begin{bmatrix} u \\ v \\ r \end{bmatrix}
$$

$$
+ \begin{bmatrix} -X_u & 0 & 0 \\ 0 & -Y_u & -Y_r \\ 0 & -N_v & -N_r \end{bmatrix} \begin{bmatrix} u \\ v \\ r \end{bmatrix} = \begin{bmatrix} \tau_u \\ 0 \\ \tau_r \end{bmatrix} \quad (11.3)
$$

$$
\begin{bmatrix} \dot{x} \\ \dot{y} \\ \dot{\psi} \end{bmatrix} = \begin{bmatrix} \cos(\psi) & -\sin(\psi) & 0 \\ \sin(\psi) & \cos(\psi) & 0 \\ 0 & 0 & 1 \end{bmatrix} \begin{bmatrix} u \\ v \\ r \end{bmatrix} \quad (11.4)
$$

where $f(v,r) = m_{23}r + m_{22}v$. The matrices are denoted M, $C(\nu)$,

D and $J(\eta)$ according to (11.1)-(11.2). M and D are both constant, positive definite matrices. The vector $\nu = [u, v, r]^T$ denotes the linear velocities in surge, sway and the angular velocity in yaw respectively. $\eta = [x, y, \psi]^T$ is the position and orientation vector and $\tau = [\tau_u, 0, \tau_r]^T$ denotes the control forces in surge and the control torque in yaw respectively.

The non-linear model for an underactuated hovercraft is obtained by simplifying the surface vessel model presented above (see [85] and [84]). We have neglected damping, considered that the shape of the hovercraft is a disc and that the propellers are located at the center of mass as shown in Figure 11.2. In order to obtain a simple model capturing the essential non-linearities of a hovercraft, we assumed the inertia matrix in (11.3) to be diagonal and equal to the identity matrix. Moreover, we cancelled the hydrodynamic damping, which is not essential in controlling the system. The dynamic equations are then given by (see [82])

$$
\begin{aligned}
\dot{u} &= vr + \tau_u \\
\dot{v} &= -ur \\
\dot{r} &= \tau_r
\end{aligned}
\tag{11.5}
$$

where τ_u is the control force in surge and τ_r is the control torque in yaw. In the second equation of system (11.5), the right term represents Coriolis and centripetal forces.

11.2.2 Euler-Lagrange's equations

Using the same assumptions as the previous section, the Lagrangian function for the system described in Figure 11.2 is given by

$$
L = \frac{1}{2}\dot{x}^2 + \frac{1}{2}\dot{y}^2 + \frac{1}{2}\dot{\psi}^2
\tag{11.6}
$$

The corresponding equations of motion are derived using Lagrange's equations

$$
\frac{d}{dt}\left(\frac{\partial L}{\partial \dot{q}}(q, \dot{q})\right) - \frac{\partial L}{\partial q}(q, \dot{q}) = \tau
\tag{11.7}
$$

where $q = [x,\ y,\ \psi]^T$ and $\tau = [\tau_u \cos(\psi),\ \tau_u \sin(\psi),\ \tau_r]^T$. From Lagrange's equations (11.7), we therefore have

$$
\begin{aligned}
\ddot{x} &= \tau_u \cos(\psi) \\
\ddot{y} &= \tau_u \sin(\psi) \\
\ddot{\psi} &= \tau_r
\end{aligned}
$$

Let us recall the kinematics (11.4) as follows

$$
\begin{aligned}
\dot{x} &= \cos(\psi)u - \sin(\psi)v \\
\dot{y} &= \sin(\psi)u + \cos(\psi)v \\
\dot{\psi} &= r
\end{aligned}
\tag{11.8}
$$

Differentiating the above equations (11.8), we obtain

$$
\ddot{x} = -\sin(\psi)ru + \cos(\psi)\dot{u} - \cos(\psi)rv - \sin(\psi)\dot{v} \tag{11.9}
$$
$$
\ddot{y} = \cos(\psi)ru + \sin(\psi)\dot{u} - \sin(\psi)rv + \cos(\psi)\dot{v} \tag{11.10}
$$
$$
\ddot{\psi} = \tau_r
$$

Multiplying (11.9) by $\cos(\psi)$ and (11.10) by $\sin(\psi)$ and adding these two equations yields

$$
\begin{aligned}
\dot{u} &= \cos(\psi)\ddot{x} + \sin(\psi)\ddot{y} + vr \\
\dot{u} &= \tau_u + vr
\end{aligned}
\tag{11.11}
$$

Multiplying (11.9) by $\sin(\psi)$ and (11.10) by $\cos(\psi)$ and adding these two equations yields

$$
\begin{aligned}
\dot{v} &= \cos(\psi)\ddot{y} - \sin(\psi)\ddot{x} - ur \\
\dot{v} &= -ur
\end{aligned}
\tag{11.12}
$$

We finally obtain the dynamic equations (11.5)

$$
\begin{aligned}
\dot{u} &= \tau_u + vr \\
\dot{v} &= -ur \\
\dot{r} &= \tau_r
\end{aligned}
\tag{11.13}
$$

We will, in the following, consider the problem of controlling the position, not the yaw angle ψ and thus disregard the latter equation in (11.8). In order to achieve simpler polynomial kinematic equations and to eliminate ψ, we use the following coordinate transformation as in [82], which is a global diffeomorphism

$$
\begin{aligned}
z_1 &= \cos(\psi)x + \sin(\psi)y \\
z_2 &= -\sin(\psi)x + \cos(\psi)y \\
z_3 &= \psi
\end{aligned}
\tag{11.14}
$$

Differentiating z_1 and z_2 and using (11.8), we obtain

$$
\begin{aligned}
\dot{z}_1 &= u + z_2 r \\
\dot{z}_2 &= v - z_1 r
\end{aligned}
\tag{11.15}
$$

The resulting model, including the kinematics and the dynamics, is finally given by

$$
\begin{aligned}
\dot{u} &= vr + \tau_u \\
\dot{v} &= -ur \\
\dot{r} &= \tau_r \\
\dot{z}_1 &= u + z_2 r \\
\dot{z}_2 &= v - z_1 r
\end{aligned}
\tag{11.16}
$$

11.2.3 Controllability of the linearized system

Since the third equation ($\dot{r} = \tau_r$) in (11.16) is directly controllable, let us consider the linearization of the four other equations.
The system can be rewritten as follows

$$
\frac{d}{dt}
\begin{bmatrix} u \\ v \\ z_1 \\ z_2 \end{bmatrix}
=
\begin{bmatrix}
0 & r & 0 & 0 \\
-r & 0 & 0 & 0 \\
1 & 0 & 0 & r \\
0 & 1 & -r & 0
\end{bmatrix}
\begin{bmatrix} u \\ v \\ z_1 \\ z_2 \end{bmatrix}
+
\begin{bmatrix} 1 \\ 0 \\ 0 \\ 0 \end{bmatrix}
\tau_u = AX + B\tau_u
$$

We then have

$$
B =
\begin{bmatrix} 1 \\ 0 \\ 0 \\ 0 \end{bmatrix}
\quad
AB =
\begin{bmatrix} 0 \\ -r \\ 1 \\ 0 \end{bmatrix}
\quad
A^2 B =
\begin{bmatrix} -r^2 \\ 0 \\ 0 \\ -2r \end{bmatrix}
\quad
A^3 B =
\begin{bmatrix} 0 \\ r^3 \\ -3r^2 \\ 0 \end{bmatrix}
$$

and $\det\left(B|AB|A^2 B|A^3 B\right) = 4r^4$.

Therefore, the linearized system is controllable if $r \neq 0$. A very simple control strategy can be obtained by fixing r to a constant different from zero and computing a linear controller for the input τ_u. This controller will exponentially stabilize (u, v, z_1, z_2) to the origin but r will not converge to zero.

Furthermore, if r is time-varying, we could use the Silverman's criterion to check the controllability of the system, i.e.

$$\text{rank } C(t) = \left[b(t), (A(t) - \frac{d}{dt})b(t), ..., (A(t) - \frac{d}{dt})^{n-1}b(t) \right] = 4 \quad (11.17)$$

In our case, $\det(C(t)) = 4r^4$. Therefore, if r is time-varying, the system is controllable at all time if $r(t) \neq 0$, $\forall t$.

11.3 Stabilizing control law for the velocity

The dynamics of the system are given as follows

$$\begin{aligned}
\dot{u} &= vr + \tau_u \\
\dot{v} &= -ur \\
\dot{r} &= \tau_r
\end{aligned} \quad (11.18)$$

The objective is to stop the hovercraft, i.e. to control "the state vector $[u \; v \; r]^T$" with the two inputs "τ_u and τ_r". τ_u and τ_r are the surge control force and the yaw control torque provided by the main propellers.

We propose the control law

$$\begin{aligned}
\tau_u &= -k_u u && (11.19) \\
\tau_r &= -ur - k_r(r - v) && (11.20)
\end{aligned}$$

where k_u and k_r are strictly positive constants. Consider the candidate Lyapunov function

$$V_1(u, v, r) = \frac{1}{2}u^2 + \frac{1}{2}v^2 + \frac{1}{2}(r - r_d)^2 \quad (11.21)$$

with $r_d = v$. The time derivative of V_1 is then

$$\begin{aligned}
\dot{V}_1 &= u(vr + \tau_u) - uvr + (r - v)(\tau_r + ur) \\
&= -k_u u^2 - k_r(r - v)^2
\end{aligned}$$

Using LaSalle's invariance principle, we consider the set $\Omega = \left\{(u,v,r) : \dot{V}_1(u,v,r) = 0\right\} = \{(u,v,r) : u = 0, r = v\}$. From (11.19), we see that $\tau_u = 0$ in Ω and from (11.18) this implies $r = v$ has to be zero to stay in Ω. Thus Ω contains no trajectory of (11.18) other than the trivial trajectory, and the continuous control law in (11.19) and (11.20) globally and asymptotically stabilizes the origin of the state $[u\ v\ r]^T$.

11.4 Stabilization of the position

11.4.1 First approach

In this section, we will develop a control law for positioning the hovercraft using the surge and angular velocities u and r as virtual control inputs. The model in (11.16) reduces to

$$
\begin{aligned}
\dot{z}_1 &= u + z_2 r \\
\dot{z}_2 &= v - z_1 r \\
\dot{v} &= -ur
\end{aligned}
\tag{11.22}
$$

Note that the above system satisfies Brockett's condition while it would not if we have added the equation for the course angle: $\dot{\psi} = r$. Consider the following candidate Lyapunov function

$$
V_2 = \frac{1}{2}z_1^2 + \frac{1}{2}z_2^2 + \frac{1}{2}v^2
\tag{11.23}
$$

Then

$$
\dot{V}_2 = z_1 u + z_2 v - uvr = z_1 u + v(z_2 - ur)
$$

We propose

$$
\left\{
\begin{aligned}
ur &= z_2 + v \\
u &= -\text{sign}(z_1)\phi
\end{aligned}
\right.
\tag{11.24}
$$

where $\text{sign}(0) = 1$ and ϕ is a positive definite function defined by

$$
\phi = \left[\frac{1}{2}\left(z_1^2 + z_2^2 + v^2\right)\right]^{\frac{1}{4}}
\tag{11.25}
$$

The resulting control input r is

$$r = \frac{z_2 + v}{-\text{sign}(z_1)\phi} \qquad (11.26)$$

Obviously r is a discontinuous function. The time derivative of V_2 is given by

$$\dot{V}_2 = -|z_1|\phi - v^2 \qquad (11.27)$$

It follows that $\dot{V}_2$ is negative and so V_2 converges. Therefore, z_1, z_2 and v remain bounded. Note that although r in (11.26) is a discontinuous function, r is bounded on any compact set. Integrating (11.27), it follows that $\int_0^t v^2 dt$ and $\int_0^t |z_1|\phi\, dt$ are finite. From (11.22) and (11.24), it follows that $\dot{v}$ is bounded, which implies that v is uniformly continuous. Using Barbalat's lemma, it follows that $v \to 0$. Then, since $\dot{z}_2 = v - z_1 r$ is bounded, z_2 is uniformly continuous. From (11.22) and (11.24), we have $\dot{v} = -ur = -z_2 + v$, then $\dot{v}$ is uniformly continuous. It follows that $\dot{v} \to 0$, using Barbalat's lemma. Using again $\dot{v} = -ur = -z_2 + v$ and $v \to 0$, it also follows that $z_2 \to 0$. Since V_2 converges, it follows that z_1 converges to a constant $z_1(\infty)$. We will study two different cases:

- Case a: If $z_1(\infty) = 0$, the state (z_1, z_2, v) converges asymptotically to the origin and the inputs u and r converge to zero.

- Case b: If $z_1(\infty) \neq 0$ then there exists a finite time T such that

$$|z_1| > \frac{1}{2}|z_1(\infty)| \qquad \forall t \geq T$$

Therefore

$$\int_T^t |z_1|\phi\, dt \geq \frac{|z_1(\infty)|}{2} \int_T^t \phi(t)\, dt$$

Since the left hand side of the above is finite and ϕ is uniformly continuous, it follows from Barbalat's lemma that $\phi \to 0$.

Finally, we conclude that the state (z_1, z_2, v) and the inputs u and r converge asymptotically to zero.

11.4.2 Second approach

In this section, we present an alternative control scheme for achieving positioning of the hovercraft. The advantage of the control strategy proposed here is that the control inputs are smoother than those of the control proposed in the previous section. We will prove also that the state (z_1, z_2, v) and the control inputs u converge to zero. However, we will only be able to prove that r remains bounded. The main idea is to choose u and r such that (see (11.22))

$$\dot{v} + \dot{z}_2 = v - (u + z_1)\, r \triangleq -(v + z_2) \tag{11.28}$$

We propose the candidate Lyapunov function

$$V_3 = \frac{1}{2}\left(z_1^2 + z_2^2\right) + \frac{1}{4}\left(v + z_2\right)^2 \tag{11.29}$$

Differentiating (11.29) and using (11.28), it follows that (see (11.22))

$$
\begin{aligned}
\dot{V}_3 &= z_1 u + z_2 v - \frac{1}{2}\left(v + z_2\right)^2 \\
&= z_1 u - \frac{v^2}{2} - \frac{z_2^2}{2}
\end{aligned}
\tag{11.30}
$$

Considering the following control inputs u and r

$$u = -z_1 + \sqrt{\frac{v^2}{4} + \frac{z_2^2}{4}} \tag{11.31}$$

and

$$r = \frac{4v + 2z_2}{\sqrt{v^2 + z_2^2}} \tag{11.32}$$

The time derivative of V_3 becomes

$$\dot{V}_3 = -z_1^2 - \frac{1}{2}v^2 - \frac{z_2^2}{2} + z_1\sqrt{\frac{v^2}{4} + \frac{z_2^2}{4}}$$

and by completion of squares, we get

$$\dot{V}_3 \le -\frac{3}{4}z_1^2 - \frac{1}{4}v^2 - \frac{1}{4}z_2^2 \tag{11.33}$$

Since V_3 and $-\dot{V}_3$ are both positive definite and since

$$\frac{1}{2}\left(z_1^2 + v^2 + z_2^2\right) \le \frac{1}{2}z_1^2 + \frac{1}{8}v^2 + \frac{1}{4}z_2^2 \le V_3$$

and

$$V_3 \leq \frac{1}{2}z_1^2 + z_2^2 + \frac{1}{2}v^2 \leq z_1^2 + z_2^2 + v^2$$

we have thus proved that the origin of the system (11.22) is globally and exponentially stable. Moreover, u converges to zero and r is bounded ($|r| \leq 6$).

11.4.3 Third approach

We will now propose a last alternative control scheme for controlling the position of the hovercraft. This latter approach is based on the main idea (11.28) and on the candidate Lyapunov function V_3 (11.29), which we will call V_4. The advantage of the control strategy presented here is that the state (z_1, z_2, v) converges exponentially to zero, whereas the convergence is only asymptotic in Section 11.4.1. Moreover, both control inputs u and r converge to zero.

Since $\dot{v} + \dot{z}_2 \triangleq -(v + z_2)$ persists, the time derivative of V_4 remains (see (11.30))

$$\begin{aligned} \dot{V}_4 &= z_1 u + z_2 v - \frac{1}{2}(v + z_2)^2 \\ &= z_1 u - \frac{v^2}{2} - \frac{z_2^2}{2} \end{aligned} \tag{11.34}$$

We propose

$$u = -z_1 - \text{sign}(z_1)\sqrt{|2v + z_2|} \tag{11.35}$$

and

$$r = -\text{sign}(z_1(2v + z_2))\sqrt{|2v + z_2|} \tag{11.36}$$

The time derivative of V_4 becomes

$$\dot{V}_4 = -z_1^2 - |z_1|\sqrt{|2v + z_2|} - \frac{1}{2}v^2 - \frac{z_2^2}{2} \tag{11.37}$$

By completion of squares (as in Section 11.4.2), it is easy to show that this implies that the origin of the system (11.22) is globally and exponentially stable. Furthermore, u and r converge to zero.

11.5 Simulation results

In order to observe the results of the different proposed control laws, we performed simulations.

Figure 11.3 shows the results for the stabilization of the system (11.18), using the control law in (11.19) and (11.20), with $k_u = 1$ and $k_r = 1$. The initial velocities are $u(0) = 10$, $v(0) = 10$ and $r(0) = 1$. Figure 11.4 shows the simulations for the stabilization of the position of

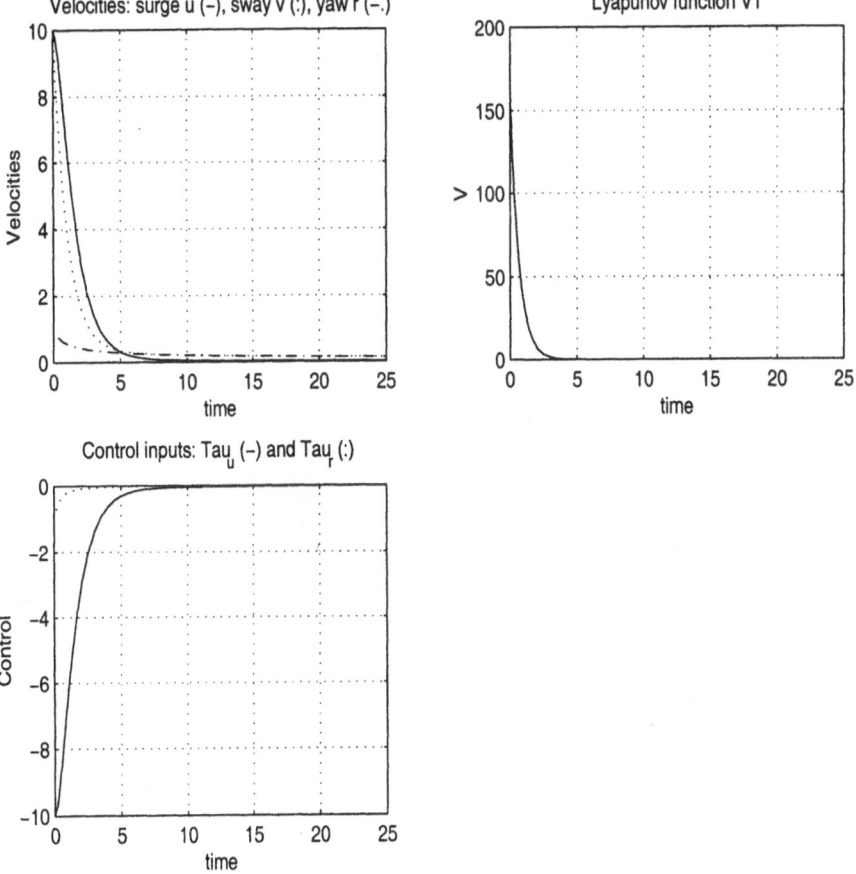

Figure 11.3: Control of the velocity using the algorithm in Section 11.3.

system (11.22) with the control law in (11.24)-(11.26). The initial positions are $z_1(0) = 0.1$, $z_2(0) = 0.1$ and $v(0) = 0$. Figure 11.5 shows the results of the control law in (11.31)-(11.32) for system (11.22), with ini-

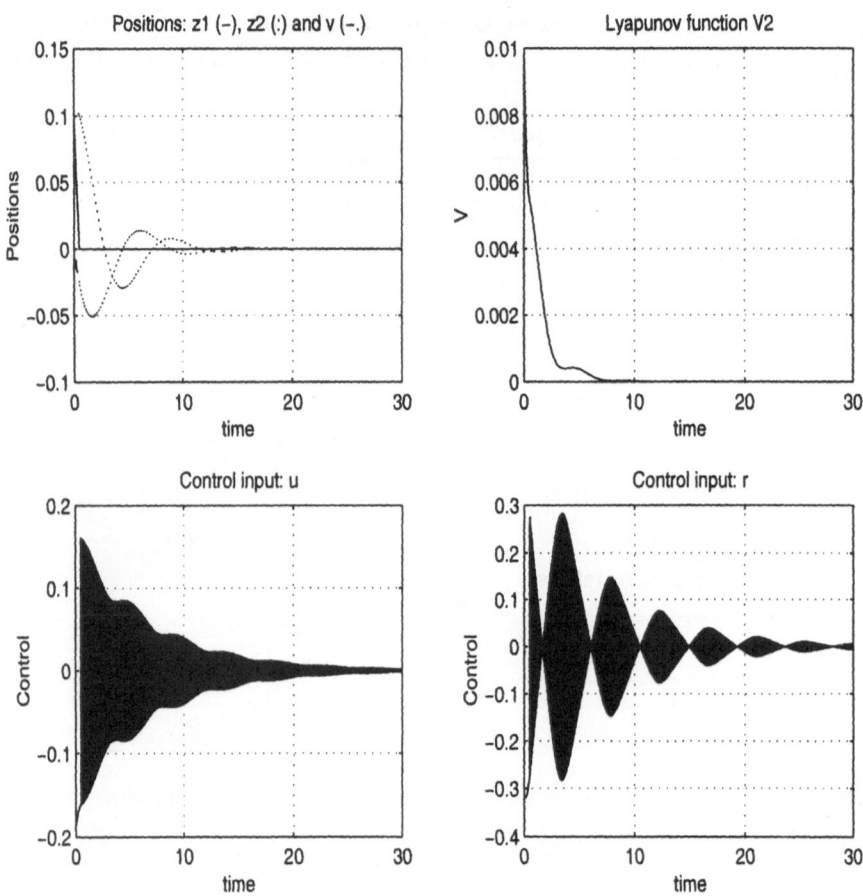

Figure 11.4: Stabilization of the position using the algorithm in Section 11.4.1: controller (11.24)-(11.26)

tial positions $z_1(0) = 10$, $z_2(0) = 10$ and $v(0) = 1$. We can choose larger initial positions, because the control does not saturate since the control law is smoother than those using the sign-function. Finally, Figure

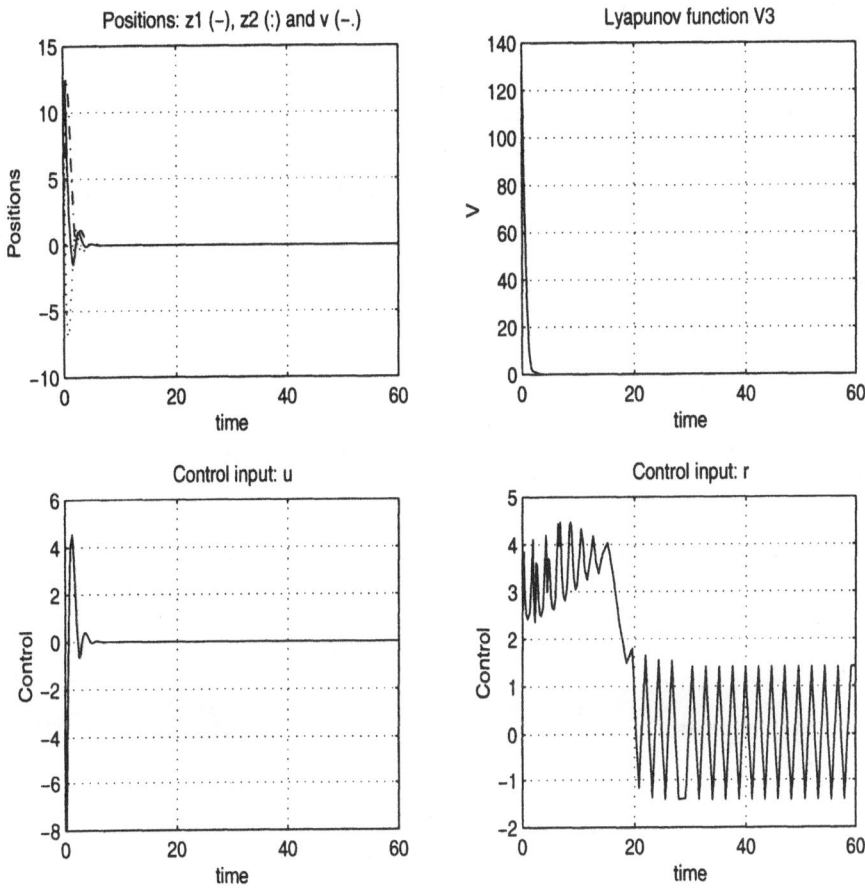

Figure 11.5: Stabilization of the position using the algorithm in Section 11.4.2: controller (11.31)-(11.32)

11.6 shows the results of the control law in (11.35)-(11.36) for system (11.22), with initial positions $z_1(0) = 0.1$, $z_2(0) = 0.1$ and $v(0) = 0$.

11.6 Conclusions

We have presented a model of an underactuated hovercraft with three degrees of freedom and two control inputs. We have proposed a control

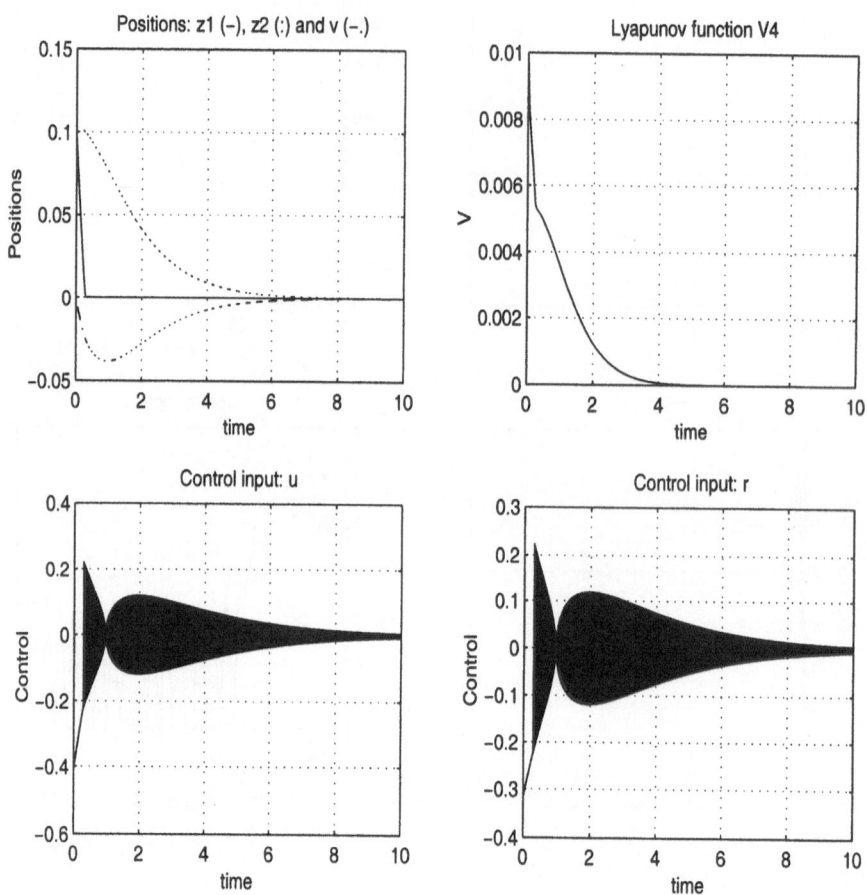

Figure 11.6: Control of the velocity using the algorithm in Section 11.4.3: (11.35)-(11.36)

scheme based on a Lyapunov approach to stabilize the surge, sway and angular velocities. We have also proposed three control strategies for positioning the vehicle using the surge and the angular velocity as virtual inputs. The three positioning controllers are discontinuous. One of the controllers is such that the origin is globally and asymptotically stable and the two inputs converge to zero. The second controller is such that the origin is globally and exponentially stable and one of the inputs (u) converges to zero while the other (r) is only proved to be bounded. The third controller is such that the origin is globally and exponentially stable and both inputs (u) and (r) converge to zero. The proposed control presents an undesired chattering behavior. Further studies are underway to better understand the control of the underactuated hovercraft model presented in this chapter. Modifications are still required to reduce the high frequency oscillations observed in simulations in order to render the controller applicable to a real system.

Chapter 12

The PVTOL aircraft

12.1 Introduction

Flight control is an essential control problem that appears in many applications such as spacecraft, aircraft and helicopters. The complete dynamics of an aircraft, taking into account aeroelastic effects, flexibility of the wings, internal dynamics of the engine and the multitude of changing variables, are quite complex and somewhat unmanageable for the purposes of control. It is also particularly interesting to consider a simplified aircraft, which has a minimum number of states and inputs but retains the main features that must be considered when designing control laws for a real aircraft. Therefore, as considered by Hauser et al. [35], we focus our study on the planar vertical take-off and landing (PVTOL) aircraft, which is a highly manoeuvrable jet aircraft.

A picture of a real vertical and short take-off and landing is shown in Figure 12.1. This is the Bell X-22A V/STOL, the last aircraft to be manufactured in Western New York. It can be seen at the Niagara Aerospace Museum and it is on loan from the National Museum of Naval Aviation.

Several methodologies for controlling such a system exist in the literature. Hauser et al. [35] in 1992 developed an approximate I-O linearization procedure, which resulted in bounded tracking and asymptotic stability for the V/STOL aircraft. In 1996, Teel [115] illustrated his central result of non-linear small gain theorem for the example of the PVTOL aircraft with input corruption. His theorem provided a formalism for analyzing the behavior of certain control systems with saturation. He established a stabilization algorithm for non-linear systems in so-called feedforward form and illustrated his result with the

Figure 12.1: The Bell X-22A V/STOL

example of the PVTOL aircraft.

In 1996 also, Martin et al. [67] proposed an extension of the result proposed by Hauser [35]. Their idea was to find a flat output for the system and to split the output tracking problem in two steps. Firstly, they designed a state tracker based on exact linearization by using the flat output and secondly, they designed a trajectory generator to feed the state tracker. They thus controlled the tracking output through the flat output. In contrast to the approximate linearization-based control method proposed by Hauser, their control scheme provided output tracking of non-minimum phase flat systems. They also took into account in the design the coupling between the rolling moment and the lateral acceleration of the aircraft (i.e. $\varepsilon \neq 0$).

A paper on controlling a PVTOL aircraft appeared in 1999, where Lin et al. [54] studied robust hovering control of the PVTOL in designing a non-linear state feedback by optimal control approach.

In 2000, Reza Olfati-Saber [79] proposed a global configuration stabilization for the VTOL aircraft with a strong input coupling using a smooth static state feedback. This approach follows the ideas in [91].

The Lyapunov approach is an important stability analysis tool, since it offers robustness properties. The control strategies for the PVTOL proposed in the literature have not been based so far on the Lyapunov approach. In [115], it was suggested that the method in [72] could be used to obtain a controller for the PVTOL that is stable in the Lyapunov sense. The present chapter proves that this is indeed the case.

In this chapter, we first present a synthesis of the significant approaches that exist in the literature. We finally propose a smooth control Lyapunov function for the stabilization of the PVTOL aircraft. The construction of the proposed Lyapunov function relies on a non-linear stabilization technique called forwarding. The PVTOL aircraft model represents a particular example that illustrates well the "Lyapunov forwarding" technique. The control algorithm follows the idea developed in [115]. Indeed, the control law is such that the objective is to stabilize the altitude independently of the other variables. Then, the roll angle is stabilized in such a way that it remains within a pre-specified range. This avoids singularities in the altitude control part. Finally, the controller takes care of the distance. This strategy is also interesting because it allows satisfaction of the constraints on the altitude and the roll angle imposed by a real application. It also allows initial conditions to be handled at a distance that is far from the desired position, without affecting the altitude and/or the roll angle of the aircraft. Note that the proposed method has been presented in [20].

The chapter is organized as follows. In Section 12.2, the equations of motion for the PVTOL are recalled. The input-output linearization of the PVTOL aircraft system is presented in Section 12.3. A stabilization algorithm based on the non-linear small gain theorem is presented in Section 12.5. In Section 12.6, we develop our stabilizing control law based on the forwarding technique for the PVTOL aircraft. Simulations for the proposed control law are presented in Section 12.7. Conclusions are finally given in Section 12.8.

12.2   The PVTOL aircraft model

The equations of motion for the PVTOL aircraft are given by (see [35])

$$
\begin{aligned}
\ddot{x} &= -\sin(\theta)u_1 + \varepsilon \cos(\theta)u_2 \\
\ddot{y} &= \cos(\theta)u_1 + \varepsilon \sin(\theta)u_2 - 1 \\
\ddot{\theta} &= u_2
\end{aligned}
\tag{12.1}
$$

where x, y denote the horizontal and the vertical position of the aircraft center of mass and θ is the roll angle that the aircraft makes with the horizon. The control inputs u_1 and u_2 are the thrust (directed out of the bottom of the aircraft) and the angular acceleration (rolling moment). The parameter ε is a small coefficient that characterizes the coupling between the rolling moment and the lateral acceleration of the aircraft.

The coefficient "−1" is the normalized gravitational acceleration. The following Figure 12.2 provides a representation of the system. In the

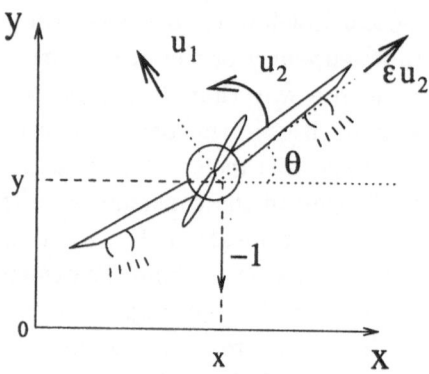

Figure 12.2: The PVTOL aircraft (front view)

present chapter, we will consider a simplified model of the PVTOL aircraft system, i.e. with $\varepsilon = 0$. Indeed, we propose to control the system as if there were no coupling between rolling moments and lateral acceleration. Therefore, the equations of motion of the system (12.1) become

$$
\begin{aligned}
\ddot{x} &= -\sin(\theta)u_1 \\
\ddot{y} &= \cos(\theta)u_1 - 1 \\
\ddot{\theta} &= u_2
\end{aligned}
\tag{12.2}
$$

This choice is due to the fact that the coefficient ε is very small $\varepsilon \ll 1$ and not always well-known, even we expect to see a loss of performance due to the unmodeled dynamics present in the system. Moreover, we do not want to increase the complexity of the controller design and of computations we will develop in this chapter.

12.3   Input-output linearization of the system

In this section, we briefly present the linearization algorithm proposed by Hauser [35]. Since we are interested in controlling the aircraft position, we choose x and y as the outputs to be controlled. We seek a state feedback law of the form $u = \alpha(z) + \beta(z)v$ such that

$$
\begin{aligned}
x^{(k_1)} &= v_1 \\
y^{(k_2)} &= v_2
\end{aligned}
\tag{12.3}
$$

for some integers k_1 and k_2. z denotes the entire state of the system and v is the new input. Differentiating the model system outputs, x and y, we get (from equations (12.2))

$$
\begin{bmatrix} \ddot{x} \\ \ddot{y} \end{bmatrix} = \begin{bmatrix} -\sin\theta & 0 \\ \cos\theta & 0 \end{bmatrix} \begin{bmatrix} u_1 \\ u_2 \end{bmatrix} + \begin{bmatrix} 0 \\ -1 \end{bmatrix}
\tag{12.4}
$$

The matrix multiplying the control inputs u_1 and u_2 is singular, which implies that there is no static state feedback that will linearize (12.2). Since u_2 comes into the system (12.2) through $\ddot{\theta}$, we have to differentiate (12.4) at least two more times. Let us consider u_1 and $\dot{u}_1$ as states and $\ddot{u}_1$ as our new input. Differentiating (12.4) twice again gives us

$$
\begin{bmatrix} x^{(4)} \\ y^{(4)} \end{bmatrix} = \begin{bmatrix} -\sin\theta & -\cos(\theta)u_1 \\ \cos\theta & -\sin(\theta)u_1 \end{bmatrix} \begin{bmatrix} \ddot{u}_1 \\ u_2 \end{bmatrix}
$$
$$
+ \begin{bmatrix} \sin(\theta)\dot{\theta}^2 u_1 - 2\cos(\theta)\dot{\theta}\dot{u}_1 \\ -\cos(\theta)\dot{\theta}^2 u_1 - 2\sin(\theta)\dot{\theta}\dot{u}_1 \end{bmatrix}
\tag{12.5}
$$

The matrix multiplying our new inputs $(\ddot{u}_1, u_2)^T$ has a determinant equal to u_1 and therefore is invertible as long as the thrust u_1 is different from zero. Note that this fact agrees well with the intuition, since no amount of rolling will affect the motion of the PVTOL aircraft if there is no thrust to effect an acceleration.

The following dynamic state feedback law can then be applied

$$
\begin{aligned}
\begin{bmatrix} \ddot{u}_1 \\ u_2 \end{bmatrix} &= \begin{bmatrix} -\sin\theta & \cos\theta \\ -\dfrac{\cos\theta}{u_1} & -\dfrac{\sin\theta}{u_1} \end{bmatrix} \\
&\quad \left(\begin{bmatrix} -\sin(\theta)\dot{\theta}^2 u_1 + 2\cos(\theta)\dot{\theta}\dot{u}_1 \\ \cos(\theta)\dot{\theta}^2 u_1 + 2\sin(\theta)\dot{\theta}\dot{u}_1 \end{bmatrix} + \begin{bmatrix} v_1 \\ v_2 \end{bmatrix} \right) \\
&= \begin{bmatrix} \dot{\theta}^2 u_1 \\ -\dfrac{2\dot{\theta}\dot{u}_1}{u_1} \end{bmatrix} + \begin{bmatrix} -\sin\theta & \cos\theta \\ -\dfrac{\cos\theta}{u_1} & -\dfrac{\sin\theta}{u_1} \end{bmatrix} \begin{bmatrix} v_1 \\ v_2 \end{bmatrix}
\end{aligned}
\tag{12.6}
$$

which results in the linearized system

$$
\begin{bmatrix} x^{(4)} \\ y^{(4)} \end{bmatrix} = \begin{bmatrix} v_1 \\ v_2 \end{bmatrix}
\tag{12.7}
$$

We may choose (v_1, v_2) such that the outputs (x, y) will track a desired trajectory and such that we guarantee that the system becomes stable.

12.4 Second stabilization approach

We present in this section a backstepping approach related to the one introduced by Sepulcre et al. in [91]. The idea is to use θ as a virtual input. Define

$$
\begin{aligned}
r_1(x, \dot{x}) &= -k_1\dot{x} - k_2 x \\
r_2(y, \dot{y}) &= -k_3\dot{y} - k_4 y
\end{aligned}
\tag{12.8}
$$

with some appropriate coefficients k_i. We wish to have

$$
\ddot{x} = r_1(x, \dot{x}) \tag{12.9}
$$
$$
\ddot{y} = r_2(y, \dot{y}) \tag{12.10}
$$

If both conditions $\cos\theta \neq 0$ and $\ddot{y}_1 + 1 \neq 0$ are satisfied, system (12.2) can be rewritten as follows

$$
\tan\theta = -\frac{\ddot{x}}{\ddot{y} + 1} \tag{12.11}
$$
$$
u_1 = -\sin(\theta)\ddot{x} + \cos(\theta)(\ddot{y} + 1) \tag{12.12}
$$

If we choose u_1 and θ such that

$$
\tan\theta = -\frac{r_1(x, \dot{x})}{r_2(y, \dot{y}) + 1} \tag{12.13}
$$
$$
u_1 = -\sin(\theta)r_1(x, \dot{x}) + \cos(\theta)(r_2(y, \dot{y}) + 1) \tag{12.14}
$$

with $r_2(y, \dot{y}) + 1 \neq 0$, this will lead us to achieve (12.9) and (12.10). Define the error

$$
\delta_1 = \tan\theta + \frac{r_1}{r_2 + 1} \tag{12.15}
$$

and

$$
r = \frac{r_1}{r_2 + 1} \tag{12.16}
$$

Differentiating (12.15), we get

$$
\begin{aligned}
\dot{\delta}_1 &= (1 + \tan^2\theta)\dot{\theta} + \dot{r} \\
\ddot{\delta}_1 &= (1 + \tan^2\theta)(2\tan\theta\dot{\theta}^2 + u_2) + \ddot{r}
\end{aligned}
\tag{12.17}
$$

We can define a controller u_2, so the closed-loop system is given by

$$
\ddot{\delta}_1 = -k_1\dot{\delta}_1 - k_0\delta_1 \tag{12.18}
$$

where $s^2 + k_1 s + k_0$ is a stable polynomial. Therefore, $\delta_1 \to 0$.

12.5 Third stabilization algorithm

Teel proposed a very simple control algorithm for the PVTOL aircraft [115]. He basically linearized the equation for the altitude y and then used a linear controller with saturated input for the (x, θ) subsystem. Defining $\xi = \left[(x - x_d) \; \dot{x} \; \theta \; \dot{\theta} \right]^T$, the control law is chosen to be of the form

$$
\begin{aligned}
u_1 &= \frac{1}{\cos(\sigma_a(\theta))} [1 - \varepsilon \sin(\theta) u_2 - (y - y_d) - 2\dot{y}] \\
u_2 &= -\theta - 2\dot{\theta} + \sigma_1(f_1^T \xi + \sigma_2(f_2^T \xi))
\end{aligned}
\tag{12.19}
$$

where $\sigma_a(s) = \text{sign}(s) \min\{|s|, a\}$ and $a < \frac{\pi}{2}$, f_1 and f_2 are constant vectors. σ_1 and σ_2 are

$$
\sigma_i(s) = \begin{cases} s & \text{for} \quad -M_i \le s \le M_i \\ \text{sign}(s) M_i & \text{for} \quad |s| > M_i \end{cases}
\tag{12.20}
$$

The controller is shown to be robust to input corruption (uncertainty in ε) at least as long as $(\theta, \dot{\theta}, y - y_d, \dot{y})$ start off sufficiently small.

12.6 Forwarding control law

This section presents our control law based on the forwarding technique. Let us consider the first order equations of the system (12.1), with $u_1 = 1 + v_1$ and $\varepsilon = 0$

$$
\begin{aligned}
\dot{x}_1 &= x_2 \\
\dot{x}_2 &= -\sin(\theta) - \sin(\theta) v_1 \\
\dot{y}_1 &= y_2 \\
\dot{y}_2 &= \cos(\theta) + \cos(\theta) v_1 - 1 \\
\dot{\theta} &= \omega \\
\dot{\omega} &= u_2
\end{aligned}
\tag{12.21}
$$

where $x_1 = x$, $y_1 = y$. Let us define z as the vector of state variables; $z = (x_1, x_2, y_1, y_2, \theta, \omega)^T$. Note that v_1 and u_2 are the new inputs. The change of input u_1 to v_1 is useful for the Lyapunov function's construction. Moreover, it allows the free system (i.e. with no inputs)

to have the origin as an equilibrium point. It means that when there
are no inputs, i.e. for $v_1 = 0$ and $u_2 = 0$, the origin of the system
$(x_1, x_2, y_1, y_2, \theta, \omega) = (0, 0, 0, 0, 0, 0)$ is an equilibrium point.

Our control objective will be to stabilize the system around a desired
position $(x_1, x_2, y_1, y_2, \theta, \omega) = (0, 0, y_{1d}, 0, 0, 0)$, where y_{1d} is the desired
altitude, different from zero. Note that this position is also an equilib-
rium point for the system (12.21). Let us consider $\tilde{y}_1 = y_1 - y_{1d}$ and $\tilde{z}$ the
new vector of state variables such that $\tilde{z} = (x_1, x_2, \tilde{y}_1, y_2, \theta, \omega)^T$. Therefore, we wish to bring the system to $(x_1, x_2, \tilde{y}_1, y_2, \theta, \omega) = (0, 0, 0, 0, 0, 0)$.

In the present section, we will propose a Lyapunov function candidate for the system with $\varepsilon = 0$. Note that this simplification only
eliminates complex terms in the Lyapunov function construction and in
the control law. The same construction could be done for the overall
system (i.e. with $\varepsilon \neq 0$). However, to avoid increasing the complexity
of the controller design, we will consider $\varepsilon = 0$. Note that this choice is
justified since this small coefficient is often not well-known. Moreover,
we will show that our control law performs well even when we include
the term due to this coefficient in the system equations.

Several steps are necessary for the construction technique, the estab-
lishment of the Lyapunov function candidate and the resulting control
law. In the next sections, we will develop these steps.

12.6.1   First step: a Lyapunov function for the altitude-angle (y, θ)-subsystem

Our first objective is to stabilize the altitude of the aircraft around
a desired altitude y_{1d}. Let us consider the (y, θ)-subsystem, with the
control input $u_2 = -\theta - \omega + v_2$. The system (12.21) becomes

$$\begin{cases} \dot{y}_1 &= y_2 \\ \dot{y}_2 &= \cos(\theta) + \cos(\theta)v_1 - 1 \\ \dot{\theta} &= \omega \\ \dot{\omega} &= -\theta - \omega + v_2 \end{cases} \tag{12.22}$$

Note that introducing the input v_2 leads to a stable subsystem (θ, ω)
for any bounded input signal v_2. We propose the Lyapunov function
candidate

$$\begin{aligned} V_1(\theta, \omega) &= \theta^2 + \omega^2 + \theta\omega \tag{12.23} \\ &= [\theta \; \omega] \begin{bmatrix} 1 & \frac{1}{2} \\ \frac{1}{2} & 1 \end{bmatrix} \begin{bmatrix} \theta \\ \omega \end{bmatrix} \geq 0 \end{aligned}$$

which is a positive definite and radially unbounded function.
The time derivative of V_1 along the trajectories of (12.22) is

$$
\begin{aligned}
\dot{V}_1(\theta, \omega) &= 2\theta\omega + 2\omega(-\theta - \omega + v_2) + \omega^2 + \theta(-\theta - \omega + v_2) \\
&= -\omega^2 - \theta^2 - \theta\omega + (2\omega + \theta)v_2 \\
&= -\frac{1}{2}[\theta^2 + \omega^2] - \frac{1}{2}[\theta + \omega]^2 + (2\omega + \theta)v_2 \\
&\leq -\frac{1}{2}[\theta^2 + \omega^2] + (2\omega + \theta)v_2 \quad\quad (12.24)
\end{aligned}
$$

At this point, we impose that $|v_2|$ be smaller than $\frac{1}{2}$. It follows that there exists T such that, for all $t \geq T$, $|\theta(t)| \leq \frac{\pi}{4}$. This will be explained in the following section.

12.6.2 Boundedness of $\theta(t)$

The subsystem (θ, ω) of (12.22) is as follows

$$
\begin{cases}
\dot{\theta} &= \omega \\
\dot{\omega} &= -\theta - \omega + v_2
\end{cases}
\quad\quad (12.25)
$$

Therefore, $\theta(t)$ satisfies

$$
\ddot{\theta}(t) + \dot{\theta}(t) + \theta(t) = v_2 \quad\quad (12.26)
$$

which is a second order differential equation. The transfer function derived from a standard form second order differential equation is

$$
\frac{\theta(s)}{V_2(s)} = \frac{k w_n^2}{(s^2 + 2\xi w_n s + w_n^2)} \qu\quad (12.27)
$$

where ξ is the damping ratio and w_n is the natural frequency. Here, $\xi = \frac{1}{2} < 1$ and $w_n = 1$. This means that the poles of the system are a complex conjugate pair $(s = \frac{-1 \pm \sqrt{3}i}{2})$. Note that the poles have negative real parts. The transient response is $e^{-\frac{1}{2}t} \cos(\frac{\sqrt{3}t}{2})$ and is oscillatory with frequency $w = \frac{\sqrt{3}}{2}$. The amplitude of the oscillations will decrease and the response will decay with time $e^{-\frac{1}{2}t}$. The system is stable.

We have imposed that v_2 be smaller than $\frac{1}{2}$. Looking at the standard form step response of a second order system with $\xi = \frac{1}{2}$ and $w_n = 1$, we can determine the maximum of the overshoot described by $\theta(t)$. See for example [52] (pages 119-120) for additional justifications. So, by applying a step input $v_2 = \frac{1}{2}$, the response of the system $\theta(t)$ is stable and therefore there exists T such that, for all $t \geq T$, $|\theta(t)| \leq \frac{\pi}{4}$.

Remark 12.1 *Note that the subsystem* (θ, ω) *is exponentially stable, given that* $|v_2| < \frac{1}{2}$. *It follows that* $|\theta(t)| < \frac{\pi}{4}$ *for some finite* $t \geq T$. *In the sequel, the results will hold for* $t \geq T$. ∎

Let us define $\tilde{y}_1 = y_1 - y_{1d}$ and the control input

$$v_1 = \frac{1}{\cos(\theta)}[1 - \cos(\theta) - \tilde{y}_1 - y_2] \tag{12.28}$$

where $\cos(\theta) \neq 0$ for $t \geq T$ in view of the constraint $|v_2| \leq \frac{1}{2}$. We assume that the system does not exhibit finite escape time during the interval $t \in [0, T)$. The system (12.22) becomes

$$\begin{cases} \dot{y}_1 &= y_2 \\ \dot{y}_2 &= -\tilde{y}_1 - y_2 \\ \dot{\theta} &= \omega \\ \dot{\omega} &= -\theta - \omega + v_2 \end{cases} \tag{12.29}$$

Note that both subsystems $(\tilde{y}_1, y_2)$ and (θ, ω) are stable for $|v_2| \leq \frac{1}{2}$. Introducing (12.28) into (12.21), we obtain

$$\begin{cases} \dot{x}_1 &= x_2 \\ \dot{x}_2 &= -\sin(\theta) + \frac{\sin(\theta)}{\cos(\theta)}[-1 + \cos(\theta)] + \frac{\sin(\theta)}{\cos(\theta)}[\tilde{y}_1 + y_2] \\ \dot{y}_1 &= y_2 \\ \dot{y}_2 &= -\tilde{y}_1 - y_2 \\ \dot{\theta} &= \omega \\ \dot{\omega} &= -\theta - \omega + v_2 \end{cases} \tag{12.30}$$

12.6.3   Second step: forwarding design

We will now use v_2 in (12.30) to control the variable x_2. Note that no control input appears in the right hand side of the second equation in (12.30). Therefore, we will control x_2 indirectly by introducing a new variable $\xi = x_2 - \omega - \theta$, which we will be able to control directly with v_2. This strategy follows the forwarding design technique. The key feature of the forwarding design is to exploit the "upper-triangular" configuration of the system to develop a "bottom-up" recursive procedure. See [72] for more details.

The time derivative of ξ satisfies

$$\dot{\xi} = -v_2 + \theta - \sin(\theta) + \frac{\sin(\theta)}{\cos(\theta)}[-1 + \cos(\theta)] + \frac{\sin(\theta)}{\cos(\theta)}[\tilde{y}_1 + y_2] \tag{12.31}$$

and

$$\sqrt{\xi^2 + 1} = \frac{\xi\dot{\xi}}{\sqrt{\xi^2 + 1}} \tag{12.32}$$

Since

$$\frac{|\xi|}{\sqrt{\xi^2 + 1}} \leq 1 \tag{12.33}$$

we have from (12.31) and (12.32)

$$\sqrt{\xi^2 + 1} \leq -\frac{\xi}{\sqrt{\xi^2 + 1}}v_2 + |\theta - \sin(\theta)| + \frac{|\sin(\theta)|}{|\cos(\theta)|}| - 1 + \cos(\theta)| \\ + \frac{|\sin(\theta)|}{|\cos(\theta)|}|\tilde{y}_1 + y_2| \tag{12.34}$$

Recall that $|\theta(t)| \leq \frac{\pi}{4}$, then $|\tan\theta(t)| \leq 1$. Note that if we define $g(\theta) = \pm(1 - \cos\theta) - \frac{\theta^2}{2}$, then $g'(\theta) = \pm\sin\theta - \theta$ and $g''(\theta) = \pm\cos\theta - 1 \leq 0$. Therefore, $g(\theta)$ has a maximum at $\theta = 0$, which means that $g(\theta) \leq 0$, i.e.

$$|-1 + \cos(\theta)| \leq \frac{\theta^2}{2} \tag{12.35}$$

Similarly, we obtain

$$|\theta - \sin(\theta)| \leq \frac{\theta^2}{2} \tag{12.36}$$

For $0 \leq \theta \leq \frac{\pi}{4}$, we have $\tan\theta \geq 0$ and then $\tan\theta + \sqrt{2}\theta \geq 0$. Moreover, looking at the graph of the function $f(\theta) = \tan\theta$ for $\theta \geq 0$, we remark that f cuts across the straight line $h(\theta) = \sqrt{2}\theta$ into 2 points: $\theta = 0$ and $\theta \simeq 0.91$. It turns out that the curve f is below h for $0 \leq \theta \leq 0.91$. Therefore, $(\tan\theta - \sqrt{2}\theta) \leq 0$ for $0 \leq \theta \leq 0.91$ and so $(\tan\theta + \sqrt{2}\theta)(\tan\theta - \sqrt{2}\theta) \leq 0$ for $0 \leq \theta \leq 0.91$. This implies that $\tan^2\theta \leq 2\theta^2$ for $|\theta| \leq 0.91$ and so for $|\theta| \leq \frac{\pi}{4}$.

Using this last inequality, i.e. $\tan^2 \theta \le 2\theta^2$, it follows that the last term $|\tan(\theta)||\tilde{y}_1 + y_2|$ in (12.34) satisfies

$$
\begin{aligned}
|\tan(\theta)||\tilde{y}_1 + y_2| &\le |\tan\theta||\tilde{y}_1| + |\tan\theta||y_2| \\
&\le \tfrac{1}{4}\tan^2\theta + \tilde{y}_1^2 + \tfrac{1}{4}\tan^2\theta + y_2^2 \\
&\le \tfrac{1}{2}\theta^2 + \tilde{y}_1^2 + \tfrac{1}{2}\theta^2 + y_2^2 \\
&\le \theta^2 + \tilde{y}_1^2 + y_2^2
\end{aligned}
\tag{12.37}
$$

where we have used the inequality $\quad 2ab \le a^2 + b^2 \quad \forall a, b \in \mathbb{R}$.
Finally, introducing inequalities (12.35)-(12.37) into (12.34) yields

$$
\sqrt{\dot{\xi}^2 + 1} \le -\frac{\xi}{\sqrt{\xi^2 + 1}} v_2 + 2\theta^2 + \tilde{y}_1^2 + y_2^2
\tag{12.38}
$$

We propose the following Lyapunov function candidate

$$
V_2 = \sqrt{\xi^2 + 1} - 1 + 6V_1(\theta, \omega) + 3V_1(\tilde{y}_1, y_2)
\tag{12.39}
$$

with $V_1(a, b) = a^2 + b^2 + ab$.
Note that from (12.29)

$$
\begin{aligned}
\dot{V}_1(\tilde{y}_1, y_2) &= 2\tilde{y}_1 y_2 + 2y_2(-\tilde{y}_1 - y_2) + y_2^2 + \tilde{y}_1(-\tilde{y}_1 - y_2) \\
&= -y_2^2 - \tilde{y}_1^2 - \tilde{y}_1 y_2 \\
&= -\tfrac{1}{2}\left[\tilde{y}_1^2 + y_2^2\right] - \tfrac{1}{2}\left[\tilde{y}_1 + y_2\right]^2
\end{aligned}
\tag{12.40}
$$

Note also that V_2 is a positive definite and radially unbounded function. Differentiating (12.39) and using (12.24), (12.38) and (12.40), we obtain

$$
\begin{aligned}
\dot{V}_2 &\le -\frac{\xi}{\sqrt{\xi^2+1}} v_2 + 2\theta^2 + \tilde{y}_1^2 + y_2^2 \\
&\quad -3\left[\theta^2 + \omega^2\right] + 6(2\omega + \theta)v_2 - \tfrac{3}{2}\tilde{y}_1^2 - \tfrac{3}{2}y_2^2 - \tfrac{3}{2}(\tilde{y}_1 + y_2)^2 \\
&\le \left[-\frac{\xi}{\sqrt{\xi^2+1}} + 6(2\omega + \theta)\right]v_2 - \theta^2 - 3\omega^2 - \tfrac{1}{2}\tilde{y}_1^2 - \tfrac{1}{2}y_2^2
\end{aligned}
\tag{12.41}
$$

Note that

$$
\begin{aligned}
12\omega v_2 + 6\theta v_2 &\le 12|\omega||v_2| + 6|\theta||v_2| \\
&\le 13|\omega||v_2| + 7|\theta||v_2| - |\omega||v_2| - |\theta||v_2| \\
&\le 13|\omega||v_2| + 7|\theta||v_2| - \frac{|\omega|}{\sqrt{1+\omega^2}}|v_2| - \frac{|\theta|}{\sqrt{1+\theta^2}}|v_2| \\
&\le 13|\omega||v_2| + 7|\theta||v_2| - \frac{\omega}{\sqrt{1+\omega^2}}v_2 - \frac{\theta}{\sqrt{1+\theta^2}}v_2
\end{aligned}
\tag{12.42}
$$

Since

$$13|\omega||v_2| \le \frac{13^2}{2 \times 7^2}\omega^2 + \frac{7^2}{2}v_2^2 \tag{12.43}$$

and

$$7|\theta||v_2| \le \frac{1}{2}\theta^2 + \frac{7^2}{2}v_2^2 \tag{12.44}$$

we have

$$7|\theta||v_2| + 13|\omega||v_2| \le 49v_2^2 + \frac{1}{2}\theta^2 + 2\omega^2 \tag{12.45}$$

Introducing (12.42) and (12.45) into (12.41), we then obtain (12.46)

$$\dot{V}_2 \le \left[-\frac{\xi}{\sqrt{\xi^2+1}} - \frac{\omega}{\sqrt{1+\omega^2}} - \frac{\theta}{\sqrt{1+\theta^2}} \right] v_2 + 49v_2^2$$
$$-\frac{1}{2}\theta^2 - \omega^2 - \frac{1}{2}\tilde{y}_1^2 - \frac{1}{2}y_2^2 \tag{12.46}$$

Let us define a new variable ψ

$$\psi = \frac{\xi}{\sqrt{\xi^2+1}} + \frac{\omega}{\sqrt{1+\omega^2}} + \frac{\theta}{\sqrt{1+\theta^2}} \tag{12.47}$$

We propose the control input v_2

$$v_2 = \frac{\psi}{60} + \mu_2 \tag{12.48}$$

Note that the new input μ_2 will be used to control the last coordinate x_1.

Using the control input v_2 in (12.48), we have

$$v_2^2 = \frac{\psi^2}{3600} + \frac{2}{60}\psi\mu_2 + \mu_2^2 \tag{12.49}$$

Using

$$2\left(\frac{\psi}{60}\sqrt{\frac{1}{49}}\right)\left(\sqrt{\frac{49}{1}}\mu_2\right) \le \frac{1}{49}\frac{\psi^2}{3600} + 49\mu_2^2 \tag{12.50}$$

(12.49) becomes

$$v_2^2 \le \left(1 + \frac{1}{49}\right)\frac{\psi^2}{3600} + (1 + 49)\mu_2^2 \tag{12.51}$$

and

$$49v_2^2 \leq 50\frac{\psi^2}{3600} + (49 \times 50)\mu_2^2 \qquad (12.52)$$

Therefore, introducing (12.48) and (12.52) into (12.46), we obtain (12.53)

$$\dot{V}_2 \leq -\frac{1}{360}\psi^2 - \psi\mu_2 + 2500\mu_2^2 - \frac{1}{2}\theta^2 - \omega^2 - \frac{1}{2}\tilde{y}_1^2 - \frac{1}{2}y_2^2 \qquad (12.53)$$

12.6.4 Third step: last change of coordinates

We propose the following change of coordinates, in order to include x_1 in the final Lyapunov function

$$\Phi = x_1 + 60\xi \qquad (12.54)$$

Using $\xi = x_2 - \omega - \theta$, (12.31) and (12.48), the time derivative of Φ is such that

$$
\begin{aligned}
\dot{\Phi} &= x_2 + 60\dot{\xi} \\
&= \xi + \omega + \theta + 60\left(-\frac{1}{60}\left[\frac{\xi}{\sqrt{\xi^2+1}} + \frac{\omega}{\sqrt{1+\omega^2}} + \frac{\theta}{\sqrt{1+\theta^2}}\right] - \mu_2\right. \\
&\qquad \left. +\theta - \sin(\theta) + \frac{\sin(\theta)}{\cos(\theta)}[-1 + \cos(\theta)] + \frac{\sin(\theta)}{\cos(\theta)}[\tilde{y}_1 + y_2]\right) \\
&= \xi - \frac{\xi}{\sqrt{\xi^2+1}} + \omega - \frac{\omega}{\sqrt{1+\omega^2}} + \theta - \frac{\theta}{\sqrt{1+\theta^2}} \\
&\qquad +60\left[-\mu_2 + \theta - \sin(\theta) + \frac{\sin(\theta)}{\cos(\theta)}[-1 + \cos(\theta)] + \frac{\sin(\theta)}{\cos(\theta)}[\tilde{y}_1 + y_2]\right] \\
&= -60\mu_2 + \frac{\xi^3}{\sqrt{\xi^2+1}+\xi^2+1} + \frac{\omega^3}{\sqrt{1+\omega^2}+\omega^2+1} + \frac{\theta^3}{\sqrt{1+\theta^2}+\theta^2+1} \\
&\qquad +60\left[\theta - \sin(\theta) + \frac{\sin(\theta)}{\cos(\theta)}[-1 + \cos(\theta)] + \frac{\sin(\theta)}{\cos(\theta)}[\tilde{y}_1 + y_2]\right]
\end{aligned}
$$

$$(12.55)$$

Differentiating $\sqrt{\Phi^2+1}$ and using $\frac{|\Phi|}{\sqrt{\Phi^2+1}} \leq 1$, we obtain (see also (12.32))

$$
\begin{aligned}
\sqrt{\dot{\Phi^2+1}} &\leq -60\frac{\Phi}{\sqrt{\Phi^2+1}}\mu_2 + \frac{|\xi|^3}{\sqrt{\xi^2+1}+\xi^2+1} + \frac{|\omega|^3}{\sqrt{1+\omega^2}+\omega^2+1} \\
&\qquad +\frac{|\theta|^3}{\sqrt{1+\theta^2}+\theta^2+1} + 60|\theta - \sin(\theta)| \\
&\qquad +60\left|\frac{\sin(\theta)}{\cos(\theta)}\right||-1 + \cos(\theta)| + 60\left|\frac{\sin(\theta)}{\cos(\theta)}\right||\tilde{y}_1 + y_2|
\end{aligned}
$$

$$(12.56)$$

Since

$$\frac{|\xi|}{\sqrt{\xi^2+1}} \le 1 \tag{12.57}$$

then

$$
\begin{aligned}
\frac{|\xi|^3}{\sqrt{\xi^2+1}+\xi^2+1} &\le \frac{|\xi|^3}{\xi^2+1} \\
&\le \frac{|\xi|}{\sqrt{\xi^2+1}}\frac{|\xi|^2}{\sqrt{\xi^2+1}} \\
&\le \frac{|\xi|^2}{\sqrt{\xi^2+1}}
\end{aligned}
\tag{12.58}
$$

Using the above and the same procedure used to obtain (12.38), it follows that

$$
\begin{aligned}
\sqrt{\dot{\Phi}^2+1} &\le -60\frac{\Phi}{\sqrt{\Phi^2+1}}\mu_2 + \frac{|\xi|^2}{\sqrt{\xi^2+1}} + \frac{|\omega|^2}{\sqrt{1+\omega^2}} + \frac{|\theta|^2}{\sqrt{1+\theta^2}} \\
&\quad +120\,\theta^2 + 60[\tilde{y}_1^2 + y_2^2] \\
&\le -60\frac{\Phi}{\sqrt{\Phi^2+1}}\mu_2 + \frac{|\xi|^2}{\sqrt{\xi^2+1}} + \omega^2 + \theta^2 \\
&\quad +120\,\theta^2 + 60[\tilde{y}_1^2 + y_2^2] \\
&\le -60\frac{\Phi}{\sqrt{\Phi^2+1}}\mu_2 + \frac{|\xi|^2}{\sqrt{\xi^2+1}} \\
&\quad +121[\theta^2 + \omega^2] + 60[\tilde{y}_1^2 + y_2^2]
\end{aligned}
\tag{12.59}
$$

Consider the Lyapunov function candidate

$$V_3 = \sqrt{\Phi^2+1} - 1 + k(1 + V_2)^2 - k \tag{12.60}$$

where k is a strictly positive parameter to be defined later. Then, using (12.59) and (12.53), the time derivative of V_3 satisfies

$$
\begin{aligned}
\dot{V}_3 &= \sqrt{\dot{\Phi}^2+1} + 2k(1+V_2)\dot{V}_2 \\
&\le -60\frac{\Phi}{\sqrt{\Phi^2+1}}\mu_2 + \frac{|\xi|^2}{\sqrt{\xi^2+1}} + 121[\theta^2 + \omega^2] + 60[\tilde{y}_1^2 + y_2^2] \\
&\quad -2k(1+V_2)\frac{1}{360}\psi^2 - 2k(1+V_2)\psi\mu_2 \\
&\quad +2k(1+V_2)\left[-\tfrac{1}{2}\theta^2 - \omega^2 - \tfrac{1}{2}\tilde{y}_1^2 - \tfrac{1}{2}y_2^2 + 2500\mu_2^2\right]
\end{aligned}
\tag{12.61}
$$

Since $V_1(\theta, \omega)$ and $V_1(\tilde{y}_1, y_2)$ are both positive, it follows from (12.39) that

$$
\begin{aligned}
-(1 + V_2) &= -\sqrt{\xi^2+1} - 6V_1(\theta,\omega) - 3V_1(\tilde{y}_1, y_2) \\
&\le -\sqrt{\xi^2+1}
\end{aligned}
\tag{12.62}
$$

Then, $\dot{V}_3$ in (12.61) becomes

$$
\begin{aligned}
\dot{V}_3 \;\leq\; & \left[-60\frac{\Phi}{\sqrt{\Phi^2+1}} - 2k(1+V_2)\psi \right]\mu_2 \\
& + \frac{|\xi|^2}{\sqrt{\xi^2+1}} + 121[\theta^2 + \omega^2] + 60[\tilde{y}_1^2 + y_2^2] \\
& - \frac{k}{180}\sqrt{1+\xi^2}\left[\frac{\xi}{\sqrt{\xi^2+1}} + \frac{\omega}{\sqrt{1+\omega^2}} + \frac{\theta}{\sqrt{1+\theta^2}} \right]^2 \\
& + 2k(1+V_2)\left[-\tfrac{1}{2}\theta^2 - \omega^2 - \tfrac{1}{2}\tilde{y}_1^2 - \tfrac{1}{2}y_2^2 + 2500\mu_2^2 \right]
\end{aligned}
\tag{12.63}
$$

Since

$$
\left(\frac{a}{\sqrt{4}} + \sqrt{4}b \right)^2 + \left(\frac{a}{\sqrt{4}} + \sqrt{4}c \right)^2 + (b+c)^2 \geq 0
\tag{12.64}
$$

it follows that

$$
(a+b+c)^2 \geq \frac{1}{2}a^2 - 4b^2 - 4c^2
\tag{12.65}
$$

and the next inequality (12.66) is satisfied

$$
\psi^2 \;\geq\; \frac{1}{2}\frac{\xi^2}{\xi^2+1} - 4\frac{\omega^2}{1+\omega^2} - 4\frac{\theta^2}{1+\theta^2}
\tag{12.66}
$$

On the other hand, using (12.62)

$$
\begin{aligned}
\frac{4k}{180}\sqrt{1+\xi^2}\left[\frac{\omega^2}{1+\omega^2} + \frac{\theta^2}{1+\theta^2} \right] \;&\leq\; \frac{4k}{180}\sqrt{1+\xi^2}\left[\omega^2 + \theta^2 \right] \\
&\leq\; \frac{4k}{180}(1+V_2)\left[\omega^2 + \theta^2 \right] \\
&\leq\; k(1+V_2)\left[\omega^2 + \tfrac{1}{11}\theta^2 \right]
\end{aligned}
\tag{12.67}
$$

Therefore, $\dot{V}_3$ becomes

$$
\begin{aligned}
\dot{V}_3 \;\leq\; & \left[-60\frac{\Phi}{\sqrt{\Phi^2+1}} - 2k(1+V_2)\psi \right]\mu_2 + \frac{|\xi|^2}{\sqrt{\xi^2+1}} + 121[\theta^2 + \omega^2] \\
& + 60[\tilde{y}_1^2 + y_2^2] - \frac{k}{360}\sqrt{1+\xi^2}\frac{\xi^2}{\xi^2+1} + k(1+V_2)\left[-\tfrac{10}{11}\theta^2 - \omega^2 \right. \\
& \left. - \tilde{y}_1^2 - y_2^2 + 5000\mu_2^2 \right]
\end{aligned}
\tag{12.68}
$$

Choosing $k = 363$, we obtain

$$
\begin{aligned}
\dot{V}_3 \;\leq\; & \left[-60\frac{\Phi}{\sqrt{\Phi^2+1}} - 726(1+V_2)\psi \right]\mu_2 + \frac{|\xi|^2}{\sqrt{\xi^2+1}} + 121[\theta^2 + \omega^2] \\
& + 60[\tilde{y}_1^2 + y_2^2] - \frac{363}{360}\frac{\xi^2}{\sqrt{1+\xi^2}} + 363(1+V_2)\left[-\tfrac{10}{11}\theta^2 - \omega^2 \right. \\
& \left. - \tilde{y}_1^2 - y_2^2 + 5000\mu_2^2 \right]
\end{aligned}
\tag{12.69}
$$

and finally, since $V_2 \geq 0$ and defining

$$\chi = -60\frac{\Phi}{\sqrt{\Phi^2 + 1}} - 726(1 + V_2)\psi \qquad (12.70)$$

we get

$$\dot{V}_3 \leq \chi\mu_2 + 1.9 \times 10^6(1 + V_2)\mu_2^2$$
$$-\frac{1}{120}\frac{\xi^2}{\sqrt{1+\xi^2}} + (1 + V_2)\left[-209\,\theta^2 - 242\,w^2 - 303\,\tilde{y}_1^2 - 303\,y_2^2\right] \qquad (12.71)$$

We propose the following control input μ_2, where λ is a strictly positive parameter

$$\mu_2 = -\lambda\chi \qquad (12.72)$$

Introducing μ_2 (12.72) in (12.71) and choosing $(1 - 1.9 \times 10^6(1 + V_2)\lambda) \geq \frac{1}{2}$, i.e.

$$\lambda \leq \frac{1}{2 \times 1.9 \times 10^6(1 + V_2)} \qquad (12.73)$$

we then have

$$\dot{V}_3 \leq -\frac{\lambda}{2}\chi^2 - \frac{1}{120}\frac{\xi^2}{\sqrt{1+\xi^2}}$$
$$+(1 + V_2)\left[-209\,\theta^2 - 242\,w^2 - 303\,\tilde{y}_1^2 - 303\,y_2^2\right] \qquad (12.74)$$

This implies that $\dot{V}_3(\tilde{z}) < 0$, $\forall \tilde{z} \neq 0$. Therefore, $\dot{V}_3$ is negative definite. In Section 12.6.1, we have imposed $|v_2|$ to be smaller than $\frac{1}{2}$. With (12.48), it turns out that we have to impose that $|\mu_2| < \frac{9}{20}$. Therefore, we will introduce in μ_2 (12.72) a saturation function to ensure that $|\mu_2| < \frac{9}{20}$. The saturation function for μ_2 is as follows

$$\mu_2 = -sat(\lambda\chi) \qquad (12.75)$$

in order to have $|\mu_2| < \frac{9}{20}$.

Note that introducing a saturation function is equivalent to choosing a λ satisfying (12.73) and μ_2 in (12.72) satisfying $|\mu_2| < \frac{9}{20}$. For $k = 363$, (12.60) becomes

$$V_3 = \sqrt{\Phi^2 + 1} - 1 + 363(1 + V_2)^2 - 363 \qquad (12.76)$$

with

$$
\begin{aligned}
\xi &= x_2 - \omega - \theta \\
\Phi &= x_1 + 60\xi \\
V_2 &= \sqrt{\xi^2 + 1} - 1 + 6V_1(\theta, \omega) + 3V_1(\tilde{y}_1, y_2) \\
V_1(a, b) &= a^2 + b^2 + ab
\end{aligned}
\tag{12.77}
$$

Note that V_3 is positive definite and radially unbounded. Since V_3 and $-\dot{V}_3$ are both positive definite and radially unbounded, we have thus proved that the origin of the overall system $\tilde{z} = (x_1, x_2, \tilde{y}_1, y_2, \theta, \omega)$ with control inputs as in (12.28), (12.48), (12.75) and (12.70) is asymptotically stable. Note that since the closed-loop system has a stable linearization, this implies local exponential stability of the system.

The main result is stated in the following theorem.

Theorem 12.1 *Consider the system (12.21) with $u_1 = 1 + v_1$. Taking the Lyapunov function candidate V_3 defined in (12.76) and (12.77) for $t \geq T$ (see Remark 12.1), then the origin of the closed-loop system with the control law (12.28), (12.48) and (12.75) is asymptotically stable.* ∎

12.7   Simulation results

In order to validate the results of the proposed control law based on the forwarding technique (see Section 12.6), we performed simulations. We started the PVTOL at the position $(x, y) = (200, 10)$ with $\theta = 0.5$ and asked the controller to move the PVTOL to the position $(x, y) = (0, 5)$ with $\theta = 0$. The simulation results are shown in Figures 12.3, 12.4 and 12.5. For the simulations, we chose $\lambda = \frac{1}{5000}$. Note also that x_1 converges slowly to 0 (see Figure 12.3). This is due to the remaining small degree of freedom on μ_2. Moreover, we ran simulations with the same control including in the system the terms $\epsilon = 0.1$ and $\epsilon = 1$ (see (12.1)). The results are very similar as for $\epsilon = 0$.

12.8   Conclusions

We have presented a control strategy for the PVTOL aircraft that stabilizes the state to the origin. We have constructed a Lyapunov function using the forwarding technique, in order to illustrate this technique with a well-known example and then construct a Lyapunov function for the

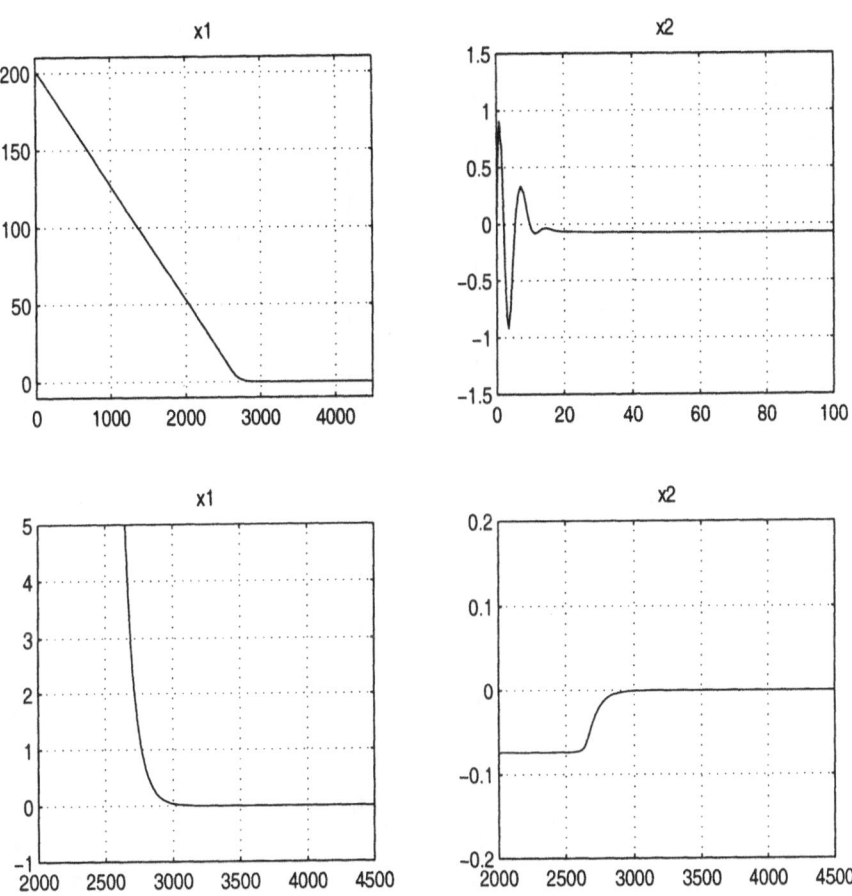

Figure 12.3: States of the system

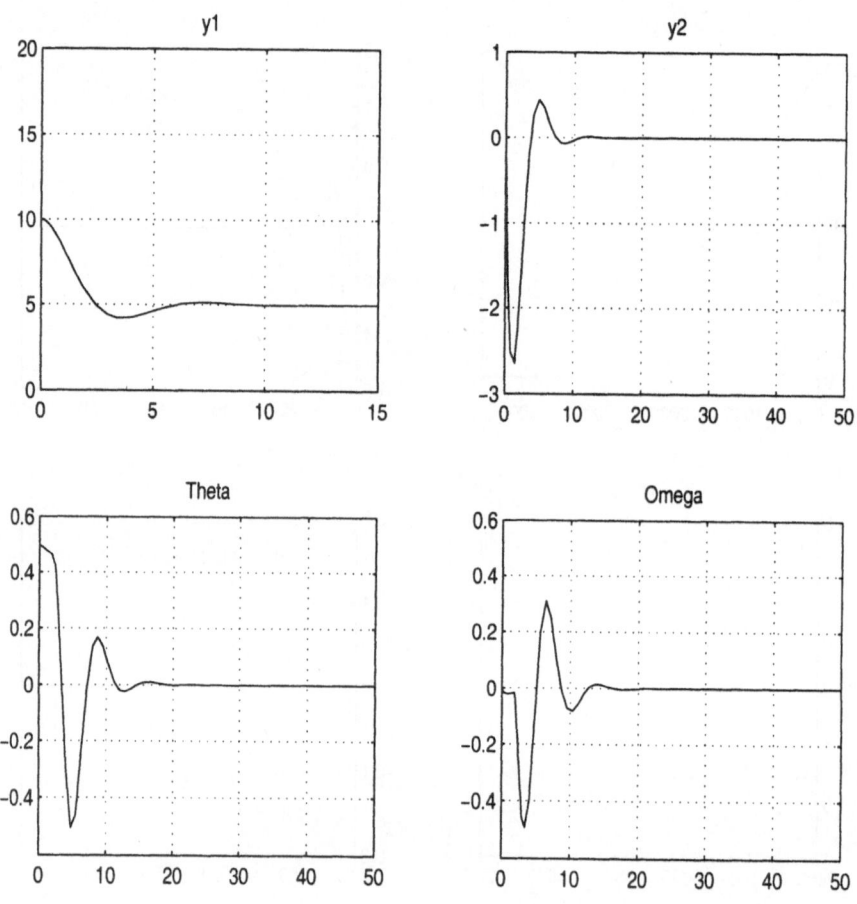

Figure 12.4: States of the system

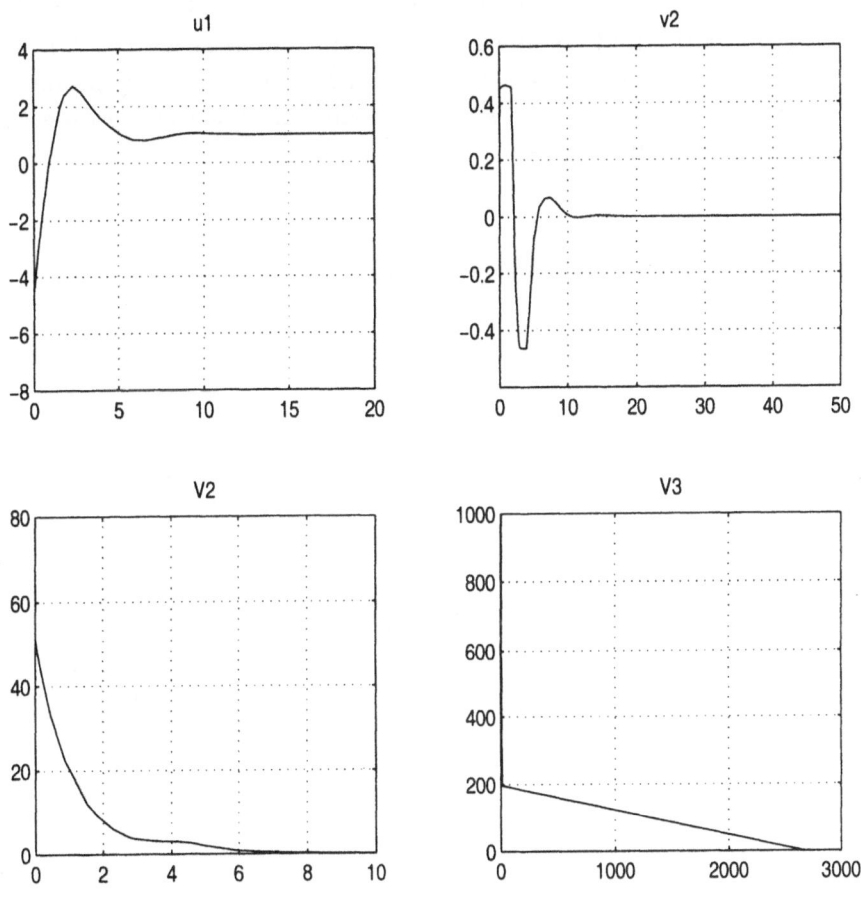

Figure 12.5: Control inputs and Lyapunov functions

PVTOL aircraft model. Good performance of the proposed control law has been tested in simulations. Compared to other controllers, the advantage lies in the fact that the control law is robust to uncertainty in ϵ (see Section 12.7). This was not the case with the approach proposed by Hauser [35]. On the other hand, the same advantage has been shown with respect to the appproach proposed by Teel [115], since our strategy is inspired by his control scheme. The controller proposed by Teel is robust to input corruption (uncertainty in ε) at least as long as $(\theta, \dot{\theta}, y - y_d, \dot{y})$ start off sufficiently small. The control strategy proposed here allows us in addition to start from an initial altitude y different from the desired altitude y_d.

Chapter 13

Helicopter on a platform

13.1 Introduction

In this chapter[*], we present a Lagrangian model of a VARIO scale model
helicopter and a passivity-based control strategy. Our global interest is
a general model (7-DOF) to be used on the autonomous forward flight
of helicopters. We present the basic idea of the 7-DOF modelling. However, in this chapter, we focus on the particular case of a reduced order
model (3-DOF) representing the scale model helicopter mounted on an
experimental platform. We note that both cases represent underactuated systems ($u \in I\!\!R^4$ for the 7-DOF model and $u \in I\!\!R^2$ for the 3-DOF
model studied in this chapter).

Vertical flight (take-off, climbing, hover, descent and landing) of the
helicopter can be analysed with the 3-DOF particular system. Although
simplified, this 3-DOF Lagrangian model presents quite interesting control challenges due to non-linearities, aerodynamical forces and underactuation. Due to the very particular dynamical and control properties
of this model, we propose a specific non-linear controller using passivity
properties.

Though the mathematical model of this system is much simpler than
that of the "free-flying" case, its dynamics will be shown to be non-
trivial (non-linear in the state, and underactuated). Some previous

---

[*]The authors of this chapter are Juan Carlos Avila Vilchis, Bernard Brogliato
and Rogelio Lozano. Juan Carlos Avila-Vilchis and Bernard Brogliato are with the
Laboratoire d'Automatique de Grenoble, France. UMR CNRS-INPG 5528. The first
author is sponsored by the UAEM (Universidad Autónoma del Estado de México).
R. Lozano is with the Laboratory Heudiasyc, UTC UMR CNRS 6599, Centre de
Recherche de Royallieu, BP 20529, 60205 Compiègne Cedex, France.

works have been developed for control problems in helicopters [48, 63, 103].

Contrary to most of the recent works in the field of non-linear control of helicopters , we incorporate the main and tail rotor dynamics in the Lagrange equations. Moreover, the control inputs are taken as the real helicopter inputs (the swash plate displacements of the main and tail rotors and the longitudinal and lateral cyclic pitch angles of the main rotor). This is shown to complicate significantly the way the input u appears in the Lagrange equations.

This chapter is organized as follows. In Section 13.2, we present some general considerations taken into account for modelling. The 3-DOF Lagrangian model of the helicopter mounted on an experimental platform is presented in Section 13.3. This model can be seen as made of two subsystems (translation and rotation). The dissipativity properties of the 3-DOF model are analyzed in Section 13.4 where one lossless operator is shown. In Section 13.5, we present a control design for the reduced order model. Section 13.6 is devoted to simulation results of the helicopter-platform system. Finally, we present some conclusions in Section 13.7.

13.2   General considerations

13.2.1   Flight modes

An experienced pilot can develop a relatively complicated take-off or free-flight (in two/three dimensions). However, helicopters often evolve in one of the following three flight modes.

Hover. When the helicopter is climbing, the pilot sets the helicopter to fly at a certain height, normally OGE (out ground effect), where the thrust of the main rotor compensates the helicopter weight mg and the vertical drag force D_{vi} produced by the wake effect (the induced velocity acting on the fuselage [1]).

Vertical Flight. This flight mode starts when the helicopter is at rest on the ground IGE (in ground effect). Then, take-off occurs and the helicopter climbs. Vertical descent precedes landing. In the

---

[1]The wake effect is a very important one that is considered in the majority of aerodynamic analysis where, for example, the pitching-up transient phenomenon produced by the induced velocity has been studied (see [114] for example).

absence of perturbations, the main rotor thrust is always vertical.

Forward Flight. We consider that this flight mode will be OGE. The thrust of the main rotor has two components. The horizontal one or traction force ensures forward flight and the vertical one keeps the helicopter at a constant height (see Figure 13.6).

In each case, the main rotor thrust orientation must allow one to compensate the pitch and roll torques that are produced on the helicopter by external perturbations. The tail rotor thrust magnitude variation will compensate the yaw torques of the same nature.

13.2.2 Aerodynamic forces and torques

In this section, we present a general panorama of the aerodynamic forces and torques computing. Our interest is to provide the reader with an idea of the 3-DOF model nature that we present in Section 13.3.

In general, the helicopter center of mass (c.m.) is not located in a plane of symmetry. Some reference systems are defined in Figure 13.1.

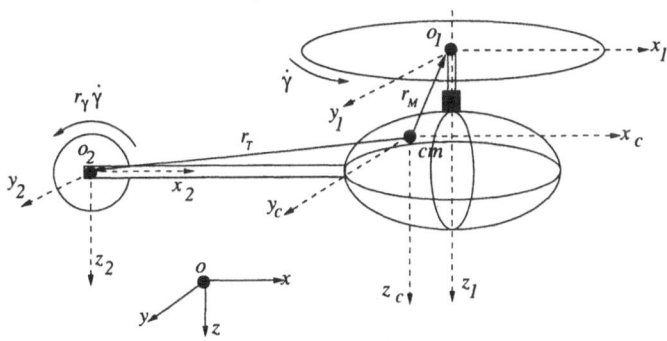

Figure 13.1: Reference systems in the helicopter

- The reference system (o, x, y, z) is an inertial one.

- The reference system (cm, x_c, y_c, z_c) is fixed at the center of mass of the helicopter and attached to its body.

- The reference system (o_1, x_1, y_1, z_1) is fixed and located at the center of the main rotor and attached to the helicopter body.

- The reference system (o_2, x_2, y_2, z_2) is fixed and located at the center of the tail rotor and attached to the helicopter body.

The most important forces and torques acting on the main and tail rotors of the helicopter are showed in Figure 13.2. In this figure, T_M is the main rotor thrust, T_T is the tail rotor thrust, C_P is the pitching moment, C_R is the rolling moment, C_Y is the yaw moment, C_M is the main rotor drag torque, C_T is the tail rotor drag torque [2], $\dot{\gamma}$ is the main rotor angular speed and r_γ is the gear ratio between the main and the tail rotors. In this work, we neglect the contributions of the horizontal and vertical stabilizers and the ground effects.

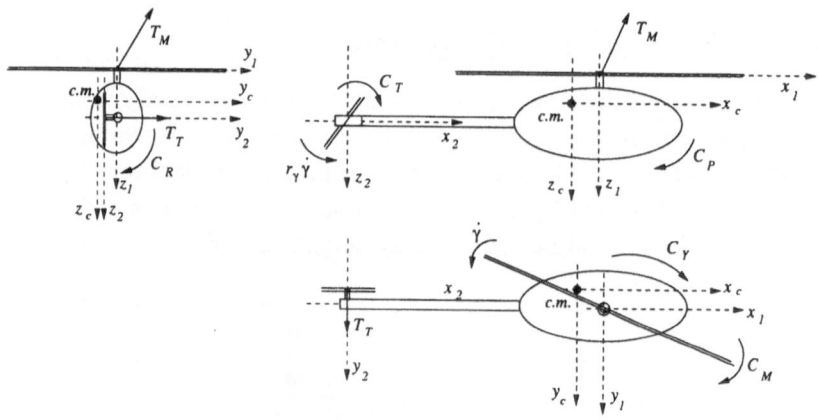

Figure 13.2: Aerodynamic forces and torques

We use the blade element method [87] to determine the magnitudes of the aerodynamic forces and torques. In Figure 13.3, we consider a main rotor blade differential element.

The lift on each blade element along the blade and around the main rotor azimuth angle γ is given by equation (13.1), where ΔL is the incremental lift (which is an infinitesimal force acting on a blade differential element), P_d is the dynamic pressure, $c_l = a\alpha$ is the lift coefficient, c is the chord and Δr_e is the incremental radial distance.

$$\Delta L = P_d c_l c \Delta r_e \qquad (13.1)$$

---

[2]All these quantities represent the magnitudes of the aerodynamic forces and torques.

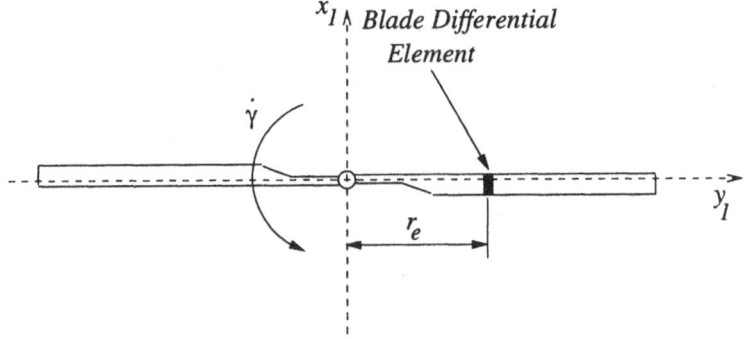

Figure 13.3: Rotor blade differential element

We can write (13.1) in terms of the blade element conditions

$$\Delta L = \frac{\rho}{2} V_T^2 a \alpha c \Delta r_e \tag{13.2}$$

where ρ is the air density, V_T is the blade element chord tangent velocity (shown in Figure 13.4), a is the slope of the lift curve and α is the angle of attack of the blade element. The lift force for a blade (L_p) and for a given azimuth angle of the main rotor is

$$L_p = \int_0^{R_M} \frac{\Delta L}{\Delta r_e} \, dr_e \tag{13.3}$$

where R_M is the radius of the main rotor. The following assumptions are taken into account in this development.

- Twist, attack, slide and flapping angles are independent of γ and of r_e.

- $\sin(\beta) \approx \beta$ where β is the flapping angle.

- $\iota_e = \arctan(\frac{V_P}{V_T}) \approx \frac{V_P}{V_T}$ where ι_e is the incidence angle.

- The flight velocity and the control inputs are independent of γ and of r_e.

The total thrust is equal to the number of blades (p) times the average lift per blade

$$T_M = \frac{p}{2\pi} \int_0^{2\pi} \int_0^{R_M} \frac{\Delta L}{\Delta r_e} \, dr_e d\gamma = \frac{\rho p a c}{4\pi} \int_0^{2\pi} \int_0^{R_M} V_T^2 \alpha \, dr_e d\gamma \tag{13.4}$$

The angle of attack α is given by equation (13.5) [87], where φ is the pitch angle defined by equation (13.6) and V_P is the velocity that is perpendicular to the blade quarter-chord line and lies in a plane that contains the rotor shaft (see Figure 13.4).

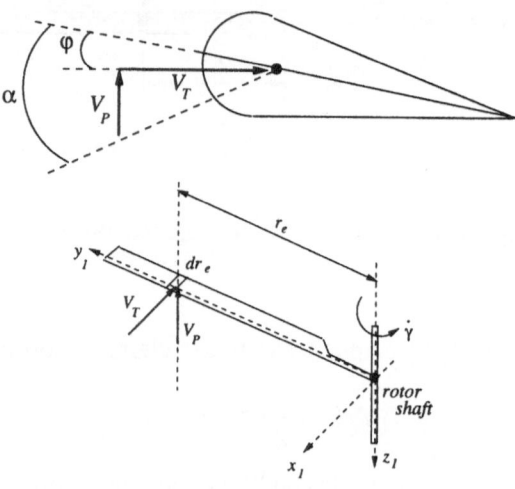

Figure 13.4: V_p and V_T velocities.

$$\alpha = \varphi + \arctan(\frac{V_P}{V_T}) \tag{13.5}$$

$$\varphi(r_e, \gamma, u_1, u_3, u_4) = \varphi_0 + \frac{r_e}{R_M}\varphi_1 - A_1 \cos(\gamma) - B_1 \sin(\gamma) \tag{13.6}$$

In (13.6), φ_0 is the average pitch at the center of rotation, φ_1 is the blade linear twist angle and A_1 and B_1 are the lateral and longitudinal cyclic pitch angles of the main rotor respectively. We can now write

$$T_M = \frac{\rho pac}{4\pi} \int_0^{2\pi} \int_0^{R_M} [V_T^2\varphi + V_T V_P]\, dr_e d\gamma \tag{13.7}$$

In the (o_1, x_1, y_1, z_1) reference system, the thrust vector is given by

$$\vec{T_M} = \begin{bmatrix} T_M \sin(u_3) \cos(u_4) \\ T_M \cos(u_3) \sin(u_4) \\ T_M \cos(u_3) \cos(u_4) \end{bmatrix} \tag{13.8}$$

with T_M given in (13.7). u_3 and u_4 are the main rotor longitudinal and lateral cyclic pitch angles respectively (i.e. $A_1 = u_4$ and $B_1 = u_3$ in (13.6)).

For the drag torque of the main rotor, we consider Figure 13.5, where the drag torque for a blade element is given by equation (13.9).

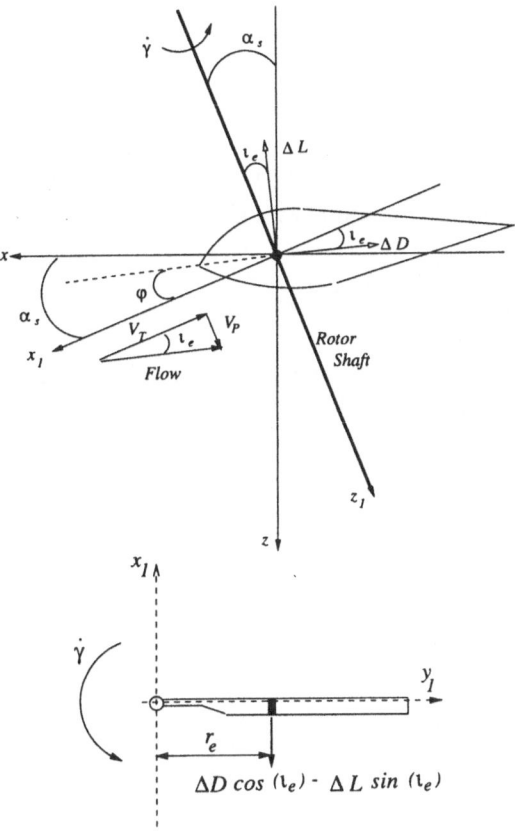

Figure 13.5: Drag torque components

$$\Delta C_M = (\Delta D \cos(\iota_e) + \Delta L \sin(\iota_e))r_e \qquad (13.9)$$

In (13.9), $\Delta D \cos(\iota_e)$ is the profile incremental drag force and $\Delta L \sin(\iota_e)$ is the induced incremental drag force due to the tilt of the lift vector. Since $\iota_e \ll 1$, the influence of $\Delta D \sin(\iota_e)$ in T_M has been neglected in (13.7).

Writing (13.9) in terms of the blade element conditions

$$\Delta C_M = (\frac{\rho}{2}c_d c V_T^2 \Delta r_e + \frac{\rho}{2}c_l c V_T^2 \Delta r_e \frac{V_P}{V_T})r_e \qquad (13.10)$$

In (13.10), c_d is the drag coefficient. We can write, taking into account that $c_l = a\alpha$

$$\Delta C_M = \frac{\rho c}{2}[c_d V_T^2 r_e + a\alpha V_T V_P r_e]\Delta r_e \qquad (13.11)$$

Taking into account equation (13.5), the number of blades and the total contribution in one revolution, we can write

$$C_M = \frac{\rho p a c}{4\pi} \int_0^{2\pi} \int_0^{R_M} [\frac{c_d}{a}(V_T^2 r_e) + (r_e V_P^2 + r_e V_T V_P \varphi)] \, dr_e d\gamma \qquad (13.12)$$

In a similar way, we can write expressions for the thrust and drag torque of the tail rotor ((13.13) and (13.14)).

$$T_T = \frac{\rho p_t a c_t}{4\pi} \int_0^{2\pi} \int_0^{R_T} [V_T^2 \varphi_q + V_T V_P]_q \, dr_e d(r_\gamma \gamma) \qquad (13.13)$$

$$C_T = \frac{\rho p_t c_t a}{4\pi} \int_0^{2\pi} \int_0^{R_T} \left[-\frac{c_{dq}}{a}((V_T)_q^2 r_e) + ((V_T)_q^2 r_e)\alpha \frac{V_P}{V_T} \right] \, dr_e d(r_\gamma \gamma) \qquad (13.14)$$

Horizontal forces or simply H-forces are not taken into account in our model. However, in Figure 13.6, we can see how these forces act on the rotors. Basic expressions for computing H-forces are

$$\Delta H_M = [(\Delta D \cos(\iota_e) + \Delta L \sin(\iota_e)) \sin(\gamma) + \Delta L \sin(\beta) \cos(\gamma)]_M \qquad (13.15)$$

and

$$\Delta H_T = [(-\Delta D \cos(\iota_e) + \Delta L \sin(\iota_e)) \sin(r_\gamma \gamma) + \Delta L \sin(\beta) \cos(r_\gamma \gamma)]_T \qquad (13.16)$$

where indices M and T concern the elements of the main and tail rotors respectively.

In this chapter, we are not interested in a detailed presentation nor in a detailed computing of all the terms involved in the 7-DOF modelling.

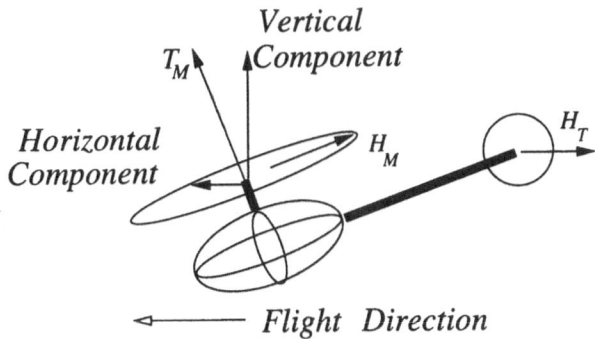

Figure 13.6: Horizontal forces

General expressions for T_M, T_T, C_M and C_T are quite complex and are not presented here. However, we can write that aerodynamic forces and torques are functions of generalized velocities and control inputs as we show below. In [7], the reader can consult detailed expressions for aerodynamic forces and torques computing and for the general forms of these forces and torques.

$$
\begin{aligned}
T_M &= T_M(\dot{x}, \dot{y}, \dot{z}, \dot{\gamma}, u_1, u_2, u_3, u_4) \\
C_M &= C_M(\dot{x}, \dot{y}, \dot{z}, \dot{\gamma}, u_1, u_2, u_3, u_4) \\
T_T &= T_T(\dot{x}, \dot{y}, \dot{z}, \dot{\gamma}, u_2) \\
C_T &= C_T(\dot{x}, \dot{y}, \dot{z}, \dot{\gamma}, u_2)
\end{aligned}
\tag{13.17}
$$

For the 7-DOF model, the control input vector is defined by $u = [u_1 \ u_2 \ u_3 \ u_4]^T$. Here, u_1, u_2 are the main and tail rotor swash plate displacements respectively. The helicopter flight velocity magnitude is given by equation (13.18) (see Figure 13.1)

$$
V^T V = \dot{x}^2 + \dot{y}^2 + \dot{z}^2
\tag{13.18}
$$

In Figures 13.7 and 13.8, we can see that for the 7-DOF model, we are considering a three-dimensional free-flight mode of the helicopter. In these figures, we take into account the real configuration that the main rotor has in a three-dimensional free-flight mode, contrary to the analysis given in [87] where the main rotor is tilted backwards (as in the case of autogiros).

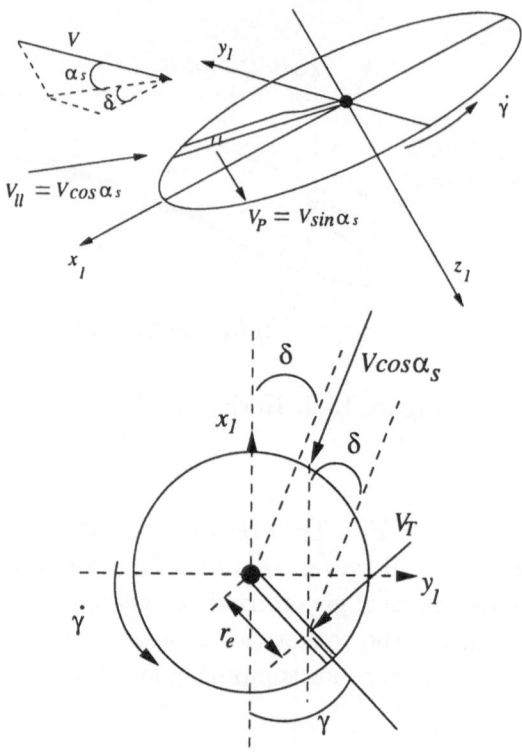

Figure 13.7: Main rotor speed components

In Figures 13.7 and 13.8, V is the flight velocity magnitude of the helicopter. The vector associated with this velocity has the opposite direction of the helicopter flight velocity, V_{ll} is the velocity that is parallel to the rotation plane and δ is the slide angle.

For the computing of aerodynamic forces and torques, only the velocity V_T that is perpendicular to the blade attack side is taken into account [87, 111]. From Figure 13.7, it is easy to write that

$$V_T = |V| \cos(\alpha_s) \sin(\gamma - \delta) + \dot{\gamma} r_e \qquad (13.19)$$

The velocity V_P is formed by several terms

$$V_P = -|V| \sin(\alpha_s) + v_{local} - r_e \dot{\beta} + |V| \cos(\alpha_s) \cos(\delta) \sin(\beta) \cos(\gamma) \qquad (13.20)$$

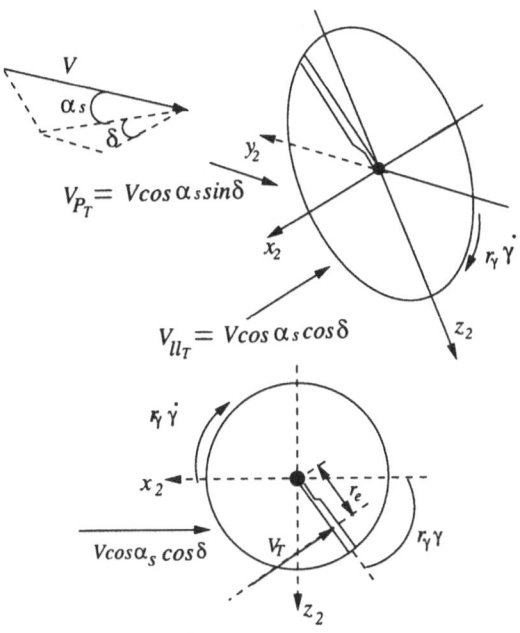

Figure 13.8: Tail rotor speed components

where $V \sin(\alpha_s)$ is the perpendicular component of the flight velocity, v_{local} is the local induced velocity [3], $r_e \dot{\beta}$ is the contribution of the vertical flapping motion and $V \cos(\alpha_s) \cos(\delta) \sin(\beta) \cos(\gamma)$ is the effect of the flight velocity component on the rotation plane of the main rotor (acting on the wing upper surface when $\gamma = 0$ and on the wing bottom surface when $\gamma = -\pi$).

When $\delta = 0$ and α_s is small, the expressions obtained in [7] for aerodynamic forces and torques become those proposed by [87] in the forward flight case. Moreover, when $V = 0$, these expressions become those of the hover mode.

13.2.3 Inertia moments and products

The main and tail rotor inertia tensors are calculated with respect to the reference system (cm, x_c, y_c, z_c). In this reference system, we denote

---

[3]Expressions to calculate the induced velocity in hover, in vertical and in forward flight can be found in [87]. In [7], one more general expression to calculate the induced velocity for a more general 3D flight mode is given.

$r_M = [x_M \ y_M \ z_M]^T$ and $r_T = [x_T \ y_T \ z_T]^T$ as the position vectors for the main and tail rotor centers respectively (see Figure 13.1).

In Figure 13.9, we represent the main and tail rotor blades. In this figure, r_e is the radial distance from the rotation center to the blade differential element dr, c and c_t are the main and tail rotor blade chords respectively, d and b are the main and tail rotor blade lengths respectively. We assume that the blade geometric form is that of a rectangular prism (dch_1 for the main rotor and $bc_t h_2$ for the tail rotor).

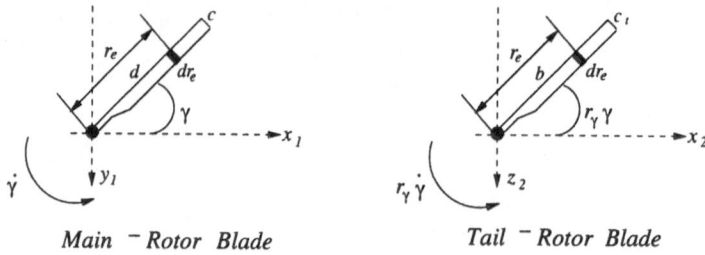

Main − Rotor Blade   Tail − Rotor Blade

Figure 13.9: Helicopter blades

We use the next classical definitions for inertia moments (13.21) and products (13.22). In these equations, we only show the main rotor notation for I_{xx} and I_{xy} with m_M the main rotor blade mass.

$$I_{xx} = \int_0^d [r^2]\, dm = \int_0^d [r^2]\, \rho dv = \int_0^d [r^2]\, \frac{m_M}{d}\, dr \qquad (13.21)$$

$$I_{xy} = \int_0^d [xy]\, dm = \int_0^d [xy]\, \rho dv = \int_0^d [xy]\, \frac{m_M}{d}\, dr \qquad (13.22)$$

In (13.21), r is the distance between the mass element and the x axis. In (13.22), x and y represent the distances between the mass element and the planes yz and zx respectively. The definitions for the rest of the inertia elements are similar.

From Figure 13.9, using (13.21) and (13.22) and for the (o_1, x_1, y_1, z_1) reference system, we can write for the main rotor (considering two blades)

$$I_{xx1} = 2\int_0^d [r_e^2 \sin^2(\gamma)\frac{m_M}{d}]\, dr_e = \frac{2}{3} m_M d^2 \sin^2(\gamma) \qquad (13.23)$$

$$I_{yy1} = 2 \int_0^d [r_e^2 \cos^2(\gamma) \frac{m_M}{d}] \, dr_e = \frac{2}{3} m_M d^2 \cos^2(\gamma) \qquad (13.24)$$

$$I_{zz1} = 2 \int_0^d [r_e^2 \frac{m_M}{d}] \, dr_e = \frac{2}{3} m_M d^2 \qquad (13.25)$$

$$I_{xy1} = -2 \int_0^d [r_e^2 \sin(\gamma) \cos(\gamma) \frac{m_M}{d}] \, dr_e = -\frac{2}{3} m_M d^2 \sin(\gamma) \cos(\gamma)$$
$$(13.26)$$

$$I_{xz1} = I_{yz1} = 0 \qquad (13.27)$$

For the tail rotor, we write for the (o_2, x_2, y_2, z_2) reference system

$$I_{xx2} = 2 \int_0^b [r_e^2 \sin^2(r_\gamma \gamma) \frac{m_T}{b}] \, dr_e = \frac{2}{3} m_T b^2 \sin^2(r_\gamma \gamma) \qquad (13.28)$$

$$I_{yy2} = 2 \int_0^b [r_e^2 \frac{m_T}{b}] \, dr_e = \frac{2}{3} m_T b^2 \qquad (13.29)$$

$$I_{zz2} = 2 \int_0^b [r_e^2 \cos^2(r_\gamma \gamma) \frac{m_T}{b}] \, dr_e = \frac{2}{3} m_T b^2 \cos^2(r_\gamma \gamma) \qquad (13.30)$$

$$I_{xy2} = I_{yz2} = 0 \qquad (13.31)$$

$$I_{xz2} = -2 \int_0^b [r_e^2 \sin(r_\gamma \gamma) \cos(r_\gamma \gamma) \frac{m_T}{b}] \, dr_e = -\frac{2}{3} m_T b^2 \sin(r_\gamma \gamma) \cos(r_\gamma \gamma)$$
$$(13.32)$$

Here, m_T is the tail rotor blade mass.

We use the parallel axes theorem (see, for example, [94]) to calculate the general inertia moments and products with respect to the (cm, x_c, y_c, z_c) reference system. With the assumption that $x_M = y_M = 0$ and $z_T = 0$, the inertia elements are simplified.

The main and tail rotor general and simplified inertia elements are given in Tables 13.1 and 13.2 respectively. Note that $D_T = x_T^2 + y_T^2$.

	General	*Simplified*
I_{xx}	$2m_M(\frac{1}{3}d^2\sin^2(\gamma) + (y_M^2 + z_M^2))$	$2m_M(\frac{1}{3}d^2\sin^2(\gamma) + z_M^2)$
I_{yy}	$2m_M(\frac{1}{3}d^2\cos^2(\gamma) + (x_M^2 + z_M^2))$	$2m_M(\frac{1}{3}d^2\cos^2(\gamma) + z_M^2)$
I_{zz}	$\frac{2}{3}m_M d^2 + 2m_M(x_M^2 + y_M^2)$	$\frac{2}{3}m_M d^2$
I_{xy}	$2m_M(-\frac{1}{3}d^2\sin(\gamma)\cos(\gamma) + x_M y_M)$	$-\frac{2}{3}m_M d^2\sin(\gamma)\cos(\gamma)$
I_{xz}	$2m_M x_M z_M$	0
I_{yz}	$2m_M y_M z_M$	0

Table 13.1: Main rotor inertia elements

	General	*Simplified*
I_{xx}	$2m_T(\frac{1}{3}b^2\sin^2(r_\gamma\gamma) + (y_T^2 + z_T^2))$	$2m_T(\frac{1}{3}b^2\sin^2(r_\gamma\gamma) + y_T^2)$
I_{yy}	$\frac{2}{3}m_T b^2 + 2m_T(x_T^2 + z_T^2)$	$\frac{2}{3}m_T b^2 + 2m_T x_T^2$
I_{zz}	$2m_T(\frac{1}{3}b^2\cos^2(r_\gamma\gamma) + D_T)$	$2m_T(\frac{1}{3}b^2\cos^2(r_\gamma\gamma) + D_T)$
I_{xy}	$2m_T x_T y_T$	$2m_T x_T y_T$
I_{xz}	$-\frac{1}{3}m_T b^2\sin(2r_\gamma\gamma) + 2m_T x_T z_T$	$-\frac{1}{3}m_T b^2\sin(2r_\gamma\gamma)$
I_{yz}	$2m_T y_T z_T$	0

Table 13.2: Tail rotor inertia elements

13.2.4 The general model

This model is based on energy considerations. We use the kinetic and the potential energies of the system. Kinetic energy is formed by four quantities, the helicopter translational energy, the fuselage rotational energy and the main and tail rotor rotational energies. Potential energy is formed by the gravitational potential energy and by the elastic potential energy associated with flapping phenomena [4].

In Section 13.3, the Lagrange equations for the 3-DOF system will be derived. This formulation will allow us to calculate the inertia and Coriolis matrices and the conservative forces vector for this particular case.

In this section, we only give the form of the general 7-DOF model. The reader is referred to [7] for details of the structure of the model elements. The general model developed for the free-flight mode of heli-

---

[4]Main rotor vertical flapping is assumed to be made up of a coning angle and of a first harmonic motion. Tail rotor flapping in neglected.

copters has the following form

$$M(q)\ddot{q} + C(q,\dot{q})\dot{q} + G(q) = J_R^T \left[D(\dot{q},u)u + A(\dot{q})u + B(\dot{q}) \right] \quad (13.33)$$

where $M \in \mathbb{R}^{7 \times 7}$ is the inertia matrix, $C \in \mathbb{R}^{7 \times 7}$ is the Coriolis matrix, $G \in \mathbb{R}^7$ is the vector of conservative forces, $q \in \mathbb{R}^7$ is the generalized coordinates vector, $J_R \in \mathbb{R}^{7 \times 7}$ is a Jacobian matrix between the generalized forces space and the external forces space applied on the helicopter and $u \in \mathbb{R}^4$ is the control input vector.

13.3 The helicopter-platform model

We consider Figure 13.10, where the VARIO helicopter mounted on an experimental platform is represented. It is important to say that in this particular case, the helicopter is in an OGE condition (platform height $\geq$ main rotor diameter). The effects of the compresed air in take-off and landing are then neglected.

In Figure 13.10, the counterbalance weight compensates the weight of the vertical column of the platform. The xyz reference system is an inertial one and the $x_1 y_1 z_1$ reference system is a body fixed frame. The model is obtained by a Lagrangian formulation. The kinetic energy T is formed by four quantities: T_t, T_{rF}, T_{rM} and T_{rT} corresponding to translational kinetic energy and rotational kinetic energies of the fuselage, of the main and of the tail rotors respectively. The potential energy is formed by the gravitational potential energy U_g and by the elastic potential energy U_b associated with the vertical flapping. In the particular case that we present here, $U_b = ka_0^2$ where k is the stiffness of the main rotor blades and a_0 is the coning angle.

The model has the following form

$$M(q)\ddot{q} + C(q,\dot{q})\dot{q} + G(q) = Q(u) \quad (13.34)$$

where $M \in \mathbb{R}^{3 \times 3}$ is the inertia matrix, $C \in \mathbb{R}^{3 \times 3}$ is the Coriolis matrix, $G \in \mathbb{R}^3$ is the vector of conservative forces, $Q = [f_z \quad \tau_z \quad \tau_\gamma]^T$ is the vector of generalized forces, $q = [z \quad \phi \quad \gamma]^T$ is the vector of generalized coordinates and $u = [u_1 \quad u_2]^T = [h_M \quad h_T]^T$ is the vector of control inputs. Here, f_z, τ_z and τ_γ are the vertical force, the yaw torque and the main rotor torque respectively. The height $z < 0$ upwards and ϕ is the yaw angle. The swash plate displacements of the main (h_M) and tail

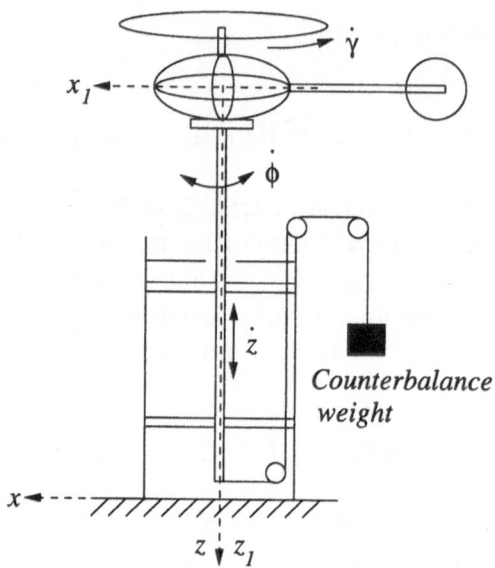

Figure 13.10: Helicopter platform

(h_T) rotors are proportional to their respective collective pitch angle as
in (13.35).

$$[\varphi_0]_k = \arctan(\frac{u_j}{bl_i}) \approx \frac{u_j}{bl_i} \tag{13.35}$$

where if $j = 1$, then $i = m$ and $k = M$, if $j = 2$ then $i = t$ and $k = T$.
Here, M and m stand for main rotor and T and t for tail rotor. In
Figure 13.11, we can see bl_i.

The components of the vector Q take the particular form (13.36) [7]:
$f_z = T_M + D_{vi}$, $\tau_z = T_T x_T$ and $\tau_\gamma = C_M + C_{mot}$. Here, C_{mot} is the
engine torque.

$$Q = \begin{bmatrix} c_8\dot{\gamma}^2 u_1 + c_9\dot{\gamma} + c_{10} \\ c_{11}\dot{\gamma}^2 u_2 \\ (c_{12}\dot{\gamma} + c_{13})u_1 + c_{14}\dot{\gamma}^2 + c_{15} \end{bmatrix} \tag{13.36}$$

Remark 13.1 *The motor dynamics are slower than those of the main
rotor. However, in scale model helicopters, there is a coupling between
the motor power and the main rotor blade collective pitch angle by the* u_1
*input as a consequence of handling conditions. In real helicopters, the
motor power is associated with an independent third input (the throttle*

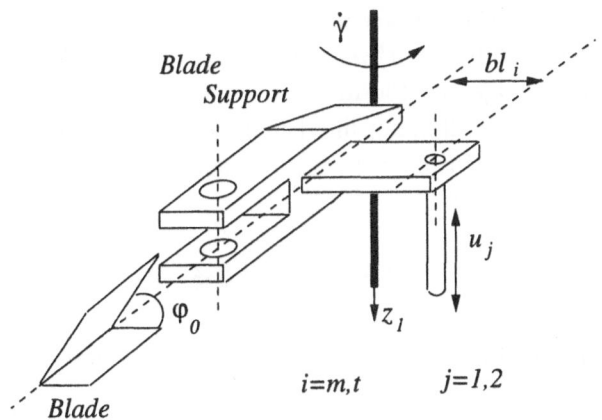

Figure 13.11: bl_i for $i = m, t$

lever) that would represent for the helicopter-platform model a completely actuated system. ∎

Remark 13.2 *The main rotor thrust T_M or $\dot{\gamma}$ are not used as inputs in (13.36) because $\ddot{\gamma}$ is very small due to the motor capabilities. Given that $\dot{u}_1$ and $\dot{u}_2$ can be larger than $\ddot{\gamma}$, u_1 and u_2 are preferred as inputs.* ∎

The Lagrangian formulation of (13.34) is as follows. The translational kinetic energy (T_t) is

$$T_t = \frac{1}{2}m\dot{z}^2 \tag{13.37}$$

where m is the helicopter mass. The rotational kinetic energies for the fuselage (T_{rF}), the main (T_{rM}) and the tail (T_{rT}) rotors are

$$T_{rF} = \frac{1}{2}I_{zzF}\dot{\phi}^2 \tag{13.38}$$

$$T_{rM} = \frac{1}{2}I_{zzM}(\dot{\phi} + \dot{\gamma})^2 \tag{13.39}$$

$$T_{rT} = \frac{1}{2}I_{zzT}\dot{\phi}^2 + \frac{1}{2}I_{yyT}\dot{\gamma}^2 r_\gamma^2 - I_{yzT}\dot{\phi}\dot{\gamma}r_\gamma \tag{13.40}$$

We denote I_{iiR} as the inertia moment with respect to the i axis and I_{ijR} the inertia product with respect to the ij axes for $R = F, M, T$. F stands for fuselage, M for main rotor and T for tail rotor. In this development, we use the simplified inertia elements of Table 13.2 and we assume that the fuselage inertia tensor is constant and diagonal. The potential energy is

$$U = -mgz + ka_0^2 \qquad (13.41)$$

where a_0 is a constant coning angle [5]. The Lagrangian function L is then given by

$$L = \tfrac{1}{2}m\dot{z}^2 + \tfrac{1}{2}I_{zzF}\dot{\phi}^2 + \tfrac{1}{2}I_{zzM}(\dot{\phi} + \dot{\gamma})^2$$
$$+ \tfrac{1}{2}I_{zzT}\dot{\phi}^2 + \tfrac{1}{2}I_{yyT}\dot{\gamma}^2 r_\gamma^2 + mgz - ka_0^2 \qquad (13.42)$$

The Lagrange equations of the helicopter motion are given by the next expressions [33]

$$\frac{d}{dt}\left[\frac{\partial L}{\partial \dot{q}_i}\right] - \frac{\partial L}{\partial q_i} = Q_i \qquad (13.43)$$

with $i = 1, 2, 3$. So we write

$$\frac{\partial L}{\partial z} = mg$$

$$\frac{\partial L}{\partial \phi} = 0 \qquad (13.44)$$

$$\frac{\partial L}{\partial \gamma} = -\tfrac{2}{3}m_T b^2 r_\gamma \sin(r_\gamma\gamma)\cos(r_\gamma\gamma)\dot{\phi}^2$$

$$\frac{\partial L}{\partial \dot{z}} = m\dot{z}$$

$$\frac{\partial L}{\partial \dot{\phi}} = I_{zzF}\dot{\phi} + I_{zzM}(\dot{\phi} + \dot{\gamma}) + 2m_T(x_T^2 + y_T^2)\dot{\phi} + \tfrac{2}{3}m_T b^2 \cos^2(r_\gamma\gamma)\dot{\phi}$$

$$\frac{\partial L}{\partial \dot{\gamma}} = I_{zzM}(\dot{\phi} + \dot{\gamma}) + I_{yyT}r_\gamma^2\dot{\gamma}$$

$$(13.45)$$

---

[5]In general, the coning angle depends on $\dot{\gamma}$ [87].

$$\frac{d}{dt}\left[\frac{\partial L}{\partial \dot{z}}\right] = m\ddot{z}$$

$$\frac{d}{dt}\left[\frac{\partial L}{\partial \dot{\phi}}\right] = I_{zzF}\ddot{\phi} + I_{zzM}(\ddot{\phi} + \ddot{\gamma}) + 2m_T(x_T^2 + y_T^2)\ddot{\phi}$$

$$+ \tfrac{2}{3}m_T b^2 \cos^2(r_\gamma\gamma)\ddot{\phi} - \tfrac{4}{3}m_T b^2 \sin(r_\gamma\gamma)\cos(r_\gamma\gamma)r_\gamma\dot{\gamma}\dot{\phi} \tag{13.46}$$

$$\frac{d}{dt}\left[\frac{\partial L}{\partial \dot{\gamma}}\right] = I_{zzM}(\ddot{\phi} + \ddot{\gamma}) + I_{yyT}r_\gamma^2\ddot{\gamma}$$

The various terms in (13.34) are obtained from (13.44), (13.45) and (13.46) as follows

$$M(q) = \begin{bmatrix} c_0 & 0 & 0 \\ 0 & c_1 + c_2\cos^2(c_3\gamma) & c_4 \\ 0 & c_4 & c_5 \end{bmatrix}$$

$$C(q,\dot{q}) = \begin{bmatrix} 0 & 0 & 0 \\ 0 & c_6\sin(2c_3\gamma)\dot{\gamma} & c_6\sin(2c_3\gamma)\dot{\phi} \\ 0 & -c_6\sin(2c_3\gamma)\dot{\phi} & 0 \end{bmatrix} \tag{13.47}$$

$$G(q) = \begin{bmatrix} c_7 \\ 0 \\ 0 \end{bmatrix}$$

We note that $c_2c_3 = -2c_6$ and that $\dot{M} - 2C \in SS(3)$ [6]. The c_i's $i = 0, ..., 7$ are the constant physical parameters given in Table 13.3. One sees that this model is made of two main coupled subsystems $S_{translation}$ and $S_{rotation}$ with states $(z, \dot{z})$ and $(\phi, \dot{\phi}, \gamma, \dot{\gamma})$ respectively. This will be used for control design.

With the assumption that the helicopter evolves at low rates of vertical velocity so that the vertical flight induced velocity (v_v) and the hover induced velocity (v_h) are approximately equal, modelling the generalized forces vector as $Q(u) = A(\dot{q})u + B(\dot{q})$ and from (13.36) we can

---

[6] $SS(n)$ represents the $n \times n$ skew-symmetric matrices set. A matrix S is said to be skew-symmetric if and only if $S + S^T = 0$.

write

$$A(\dot{q}) = \begin{bmatrix} c_8 \dot{\gamma}^2 & 0 \\ 0 & c_{11} \dot{\gamma}^2 \\ c_{12} \dot{\gamma} + c_{13} & 0 \end{bmatrix}$$

$$B(\dot{q}) = \begin{bmatrix} c_9 \dot{\gamma} + c_{10} \\ 0 \\ c_{14} \dot{\gamma}^2 + c_{15} \end{bmatrix}$$

(13.48)

The c_i's $i = 8, ..., 15$ are the constant physical parameters given in Table 13.3. The values of all the parameters are given in [7].

c_i	*Definition*	*Value*
c_0	m	$7.5\ kg$
c_1	$I_{zzF} + I_{zzM} + 2m_T(x_T^2 + y_T^2)$	$0.4305\ kg \cdot m^2$
c_2	$\frac{2}{3} m_T b^2$	$3e^{-4}\ kg \cdot m^2$
c_3	r_γ	-4.143
c_4	I_{zzM}	$0.108\ kg \cdot m^2$
c_5	$I_{zzM} + r_\gamma^2 I_{yyT}$	$0.4993\ kg \cdot m^2$
c_6	$\frac{1}{3} m_T b^2 r_\gamma$	$6.214e^{-4}\ kg \cdot m^2$
c_7	$-mg$	$-73.58\ N$
c_8	$\frac{\rho p c a R_M^3}{6 b l_m}$	$3.411\ kg$
c_9	$\frac{\rho p c a R_M^2}{4} v_h$	$0.6004\ kg \cdot m/s$
c_{10}	$\%mg$	$3.679\ N$
c_{11}	$\frac{\rho p_t c_t a R_T^3 r_\gamma^2 x_T}{6 b l_t}$	$-0.1525\ kg \cdot m$
c_{12}	$-\frac{\rho p c a R_M^3}{6 b l_m} v_h$	$12.01\ kg \cdot m/s$
c_{13}	K_{engine}	$1e^5\ N$
c_{14}	$\frac{\rho p c c_d R_M^4}{8}$	$1.205e^{-4}\ kg \cdot m^2$
c_{15}	$-\frac{5\rho p c a R_M^2}{16} v_h^2$	$-2.642\ N$

Table 13.3: 3-DOF Model physical parameters

13.4 Dissipativity properties of the 3-DOF model

The use of passivity has been at the core of the design of many feedback controllers in the past 15 years (see [56]). The interest of passivity-based controllers comes from their physical foundations (contrary to some other non-linear stabilization techniques that rely only on the state space equations structure). They also prove to provide nice experimental results [56].

The design of a passivity-based controller for (13.34) is, however, quite specific due to both the Lagrangian dynamics and the form of $Q(u)$. More precisely, the z dynamics in (13.34) plus the fact that the inputs in u are not generalized forces, precludes the dissipativity of the operators

$$O_1 : u \mapsto \dot{q}$$
$$O_2 : Q(u) \mapsto \dot{q}$$

This is a property that is crucial in the design of passivity-based controllers, which assures global tracking control of $(q(t), \dot{q}(t))$. However, the operator

$$O_3 : \bar{u} \overset{\triangle}{=} \bar{A}u + \bar{B} \mapsto \dot{\eta} \overset{\triangle}{=} [\dot{\phi} \ \dot{\gamma}]^T \tag{13.49}$$

is passive lossless, with

$$\bar{M} = \begin{bmatrix} m_{22} & m_{23} \\ m_{32} & m_{33} \end{bmatrix} \quad \bar{C} = \begin{bmatrix} c_{22} & c_{23} \\ c_{32} & 0 \end{bmatrix}$$

$$\bar{A} = \begin{bmatrix} 0 & a_{22} \\ a_{31} & 0 \end{bmatrix} \quad \bar{B} = \begin{bmatrix} 0 \\ b_3 \end{bmatrix} \tag{13.50}$$

The proof is easy by noting that $\dot{\bar{M}} - 2\bar{C}$ is skew-symmetric [56].

13.5 Control design

For feedback control purposes, we will use both the structure of the model dynamics, and the physical property of the operator $O_3 : \bar{u} \mapsto \dot{\eta}$. In the following, we assume that initially $|\dot{\gamma}(0)| \geq \delta > 0$ and $z(0) < 0$, so that $\bar{A}$ is full rank. Therefore, the control design is done as follows.

13.5.1 Passivity-based control of the rotational part

The rotational dynamics is given by (see (13.50))

$$\bar{M}(\eta)\ddot{\eta} + \bar{C}(\eta,\dot{\eta})\dot{\eta} = \bar{A}(\dot{\eta})u + \bar{B}(\dot{\eta}) \qquad (13.51)$$

It is noteworthy that these dynamics also represent the rotational part of the system when the helicopter has not taken off, i.e. when $z \equiv -L$ (see Figure 13.12).

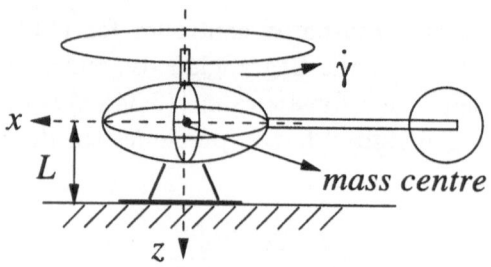

Figure 13.12: Mass center localization

Let us choose u in (13.49) such that

$$\bar{u} = \bar{M}(\ddot{\eta}_d - \lambda_1\dot{\tilde{\eta}}) + \bar{C}(\dot{\eta}_d - \lambda_1\tilde{\eta}) - \lambda_2(\dot{\tilde{\eta}} + \lambda_1\tilde{\eta}) \qquad (13.52)$$

where $\tilde{\eta} = \eta - \eta_d$, $\eta_d(t) \in C^2(\mathbb{R}^+)$ is a desired trajectory, $\lambda_1 > 0$ and $\lambda_2 > 0$. The input (13.52) is known to guarantee that $\tilde{\eta}, \dot{\tilde{\eta}}, \ddot{\tilde{\eta}} \to 0$ globally, asymptotically and exponentially [56].

13.5.2 Take-off

The basic idea is to use $\dot{\gamma}$ (in fact $\dot{\gamma}_d$) to control the first equation in (13.34), i.e.

$$c_0\ddot{z} + c_7 = c_8\dot{\gamma}^2 u_1 + c_9\dot{\gamma} + c_{10} \qquad (13.53)$$

with u_1 given by (13.52). Assuming that $\|\tilde{\eta}^{(i)}\| < \varepsilon$, $i = 0, 1, 2$, and $\dot{\phi}_d = 0$, one can approximate u_1 in (13.52) as

$$u_1^{app} = \frac{1}{a_{31}}[c_5\ddot{\gamma}_d - c_{14}\dot{\gamma}_d^2 - c_{15}] \qquad (13.54)$$

From (13.53) and (13.54), it is clear that if $\ddot{\gamma}_d$ is constant, then the altitude $z(t) = at^2 + bt + c$, $a, b, c \in \mathbb{R}$.

13.5.3 Altitude control

From (13.34), taking into account (13.36) and (13.47), we have

$$\ddot{z} = \frac{1}{c_0}[c_8\dot{\gamma}^2 u_1 + c_9\dot{\gamma} + c_{10} - c_7]$$

$$\ddot{\phi} = \frac{1}{c_1 c_5 - c_4^2 + c_2 c_5 \cos^2(c_3\gamma)}[c_5(c_{11}\dot{\gamma}^2 u_2 - 2c_6\sin(2c_3\gamma)\dot{\gamma}\dot{\phi})$$

$$-c_4((c_{12}\dot{\gamma} + c_{13})u_1 + c_6\sin(2c_3\gamma)\dot{\phi}^2 + c_{14}\dot{\gamma}^2 + c_{15})]$$

$$\ddot{\gamma} = \frac{1}{c_1 c_5 - c_4^2 + c_2 c_5 \cos^2(c_3\gamma)}[-c_4(c_{11}\dot{\gamma}^2 u_2 - 2c_6\sin(2c_3\gamma)\dot{\gamma}\dot{\phi})$$

$$+(c_1 + c_2\cos^2(c_3\gamma))((c_{12}\dot{\gamma} + c_{13})u_1 + c_6\sin(2c_3\gamma)\dot{\phi}^2$$

$$+c_{14}\dot{\gamma}^2 + c_{15})]$$

(13.55)

When the helicopter has attained a certain height ($z = -h_0$), we propose to switch the control to

$$u_1 = \frac{1}{c_8\dot{\gamma}^2}[c_7 - c_{10} - c_9\dot{\gamma} + c_0(\ddot{z}_d - \lambda_3\dot{\tilde{z}} - \lambda_4\tilde{z})]$$

$$u_2 = \frac{1}{c_5 c_{11}\dot{\gamma}^2}[(c_1 c_5 - c_4^2 + c_2 c_5\cos^2(c_3\gamma))(\ddot{\phi}_d - \lambda_5\dot{\tilde{\phi}}$$

$$-\lambda_6\tilde{\phi}) + 2c_5 c_6\sin(2c_3\gamma)\dot{\gamma}\dot{\phi} + c_4((c_{12}\dot{\gamma} + c_{13})u_1$$

$$+c_6\sin(2c_3\gamma)\dot{\phi}^2 + c_{14}\dot{\gamma}^2 + c_{15})]$$

(13.56)

Hence the closed-loop system becomes (see equation (13.55))

$$\ddot{\tilde{z}} + \lambda_3\dot{\tilde{z}} + \lambda_4\tilde{z} = 0$$

$$\ddot{\tilde{\phi}} + \lambda_5\dot{\tilde{\phi}} + \lambda_6\tilde{\phi} = 0$$

$$\ddot{\gamma} = \frac{1}{c_1 c_5 - c_4^2 + c_2 c_5\cos^2(c_3\gamma)}[-c_4(c_{11}\dot{\gamma}^2 u_2 - 2c_6\sin(2c_3\gamma)\dot{\gamma}\dot{\phi}) \quad (13.57)$$

$$+(c_1 + c_2\cos^2(c_3\gamma))((c_{12}\dot{\gamma} + c_{13})u_1 + c_6\sin(2c_3\gamma)\dot{\phi}^2$$

$$+c_{14}\dot{\gamma}^2 + c_{15})]$$

Remark 13.3 *If we consider that the motor power is associated with a different independent input u_3 (see remark 13.2), we can compensate* *by $u_1 = u_1(\dot\gamma, v_1)$ the gravity term in the first equation of (13.55), using $u_3 = u_3(u_1, v_3)$ in the last equation to compensate for the terms intro-duced by the first input and for those in the last row of vector $B(\dot q)$ and using $u_2 = u_2(\dot\gamma, v_2)$ in the second equation where v_i $i = 1, 2, 3$ are new inputs to arrive at*

$$M(q)\ddot q + C(q, \dot q)\dot q = \begin{bmatrix} v_1 \\ v_2 \\ v_3 \end{bmatrix} \tag{13.58}$$

One sees that the operator

$$O : [v_1 \; v_2 \; v_3]^T \mapsto [\dot z \; \dot\phi \; \dot\gamma]^T \tag{13.59}$$

is passive, so that a passivity-based controller can be easily designed [56].
∎

There are several crucial choices in this procedure:

i) The input in (13.52), to control the rotational dynamics of the system.

ii) $\dot\gamma_d(t)$ and a hybrid strategy, to control the translational dynamics of the system.

iii) $\eta_d(t)$ and $z_d(t)$, to comply with input saturations $u_m^1 \leq u_1 \leq 0$ and $u_m^2 \leq u_2 \leq u_M^2$. Here, $u_m^i < 0$, $i = 1, 2$ and $u_M^2 > 0$.

The technique employed to cope with input saturations is as follows: from (13.52) and assuming perfect tracking ($\tilde\eta = \dot{\tilde\eta} \equiv 0$), one sets

$$u = \bar A^{-1}(\dot\gamma_d(t))[\bar M(\eta_d)\ddot\eta_d + \bar C(\eta_d, \dot\eta_d)\dot\eta_d - \bar B(\dot\gamma_d)] \tag{13.60}$$

From this expression, one calculates off-line whether $u_m^1 \leq u_1 \leq 0$ and $u_m^2 \leq u_2 \leq u_M^2$. Then the saturations are respected, provided the initial tracking errors $\tilde\eta(0)$ and $\dot{\tilde\eta}(0)$ and the feedback gains λ_1, λ_2 are chosen to be small enough. Moreover, some numerical results show that the input may saturate during the transient without destroying the stability of the closed-loop system.

When $\ddot\gamma_d \equiv 0$, $\dot\gamma_d =$ constant, a sufficient condition to get a negative u_1 input in (13.60) is given by

$$c_6 \sin(2c_3\gamma_d)\dot\phi_d^2 + c_{14}\dot\gamma_d^2 + c_{15} \geq 0 \tag{13.61}$$

13.6 Simulation results

We performed some simulation experiments in MATLAB/SIMULINK for the 3-DOF system. Here, we present two of them. In [6], we present three different simulations with a larger number of results. We used a fixed-step ode4 Runge-Kutta solver with step 0.005. Results concerning the flying mode ($z < -L$) are not presented here, but can be easily simulated. The control of the rotational dynamics in (13.51) is in itself a challenging problem.

13.6.1 Simulation 1

We took the gain values of the control (13.52) $\lambda_1 = 8$, $\lambda_2 = 10$ and the initial conditions $\phi_0 = 0$ *rad*, $\dot{\phi}_0 = 2$ *rad/s*, $\gamma_0 = -5$ *rad* and $\dot{\gamma}_0 = -55$ *rad/s*. The helicopter was not taking-off the ground. This is a regulation problem for the desired values given below

- $\phi_d = -\frac{\pi}{4}$ *rad*

- $\dot{\phi}_d = 0$ *rad/s*

- $\ddot{\phi}_d = 0$ *rad/s^2*

- $\gamma_d = -59t$ *rad*

- $\dot{\gamma}_d = -59$ *rad/s*

- $\ddot{\gamma}_d = 0$ *rad/s^2*

13.6.2 Simulation 2

We took the same desired values as in simulation 1 but with the gain values of the control (13.52) $\lambda_1 = \lambda_2 = 1$ and the initial conditions $\phi_0 = -\frac{\pi}{4} + 0.1$ *rad*, $\dot{\phi}_0 = 1$ *rad/s*, $\gamma_0 = 0.1$ *rad* and $\dot{\gamma}_0 = -58$ *rad/s*.

In Figures 13.13 to 13.15, one sees that the input may saturate during the transient without destroying the stability of the closed-loop system. In Figures 13.16 to 13.18, one sees that decreasing the feedback gains and the initial errors allows one to respect the input saturations and to improve the transient behaviour.

13.7   Conclusions

In this chapter, we have considered the feedback control of a scale model helicopter mounted on a platform. The resulting model is a 3-DOF Lagrangian system, with two inputs. This is therefore an underactuated system. Some aerodynamical effects have been incorporated in the model to obtain the generalized torques as a function of the inputs (the swash plate displacements of the main and tail rotors) and of the main rotor angular velocity. The complete model also incorporates the transition from the constrained mode (the helicopter is at rest on the ground) to the flying mode (the helicopter is airborne).

The proposed control strategy is based on the use of non-linear controllers that assure asymptotic tracking of suitable (i.e. differentiable enough and such that the inputs do not saturate) desired trajectories.

Mechanical and aerodynamical coupling effects are taken into account in the model and in the control action. The dissipativity properties of the rotational part of the dynamics are used to partially design the state feedback control. Numerical simulations are presented to show the performance of the proposed controller.

The one-way transmission between the power shaft and the main rotor hub could be taken into account in the modelling task. Also, a first or a second order transfer function could be used to represent the motor dynamics. This will result in a more complex model. However, in this case, it is clear that the control problem will become more complex due to the additional dynamics. In this work, we have implicitly assumed that the system evolves in the bandwidth of these motor dynamics.

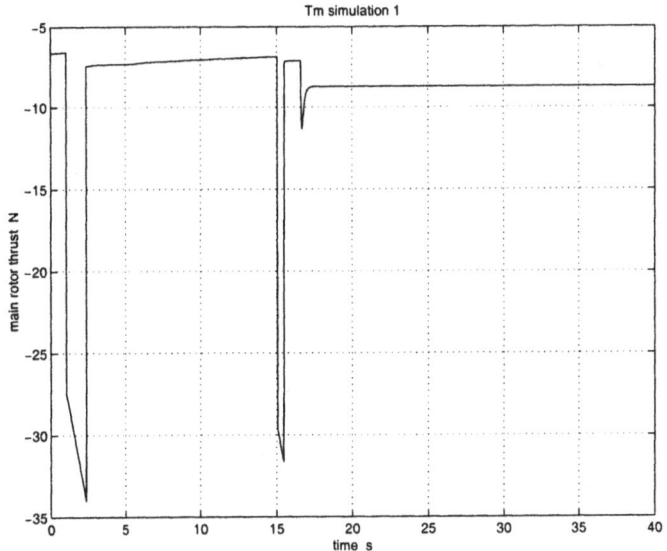

Figure 13.13: Thrust variation for simulation 1

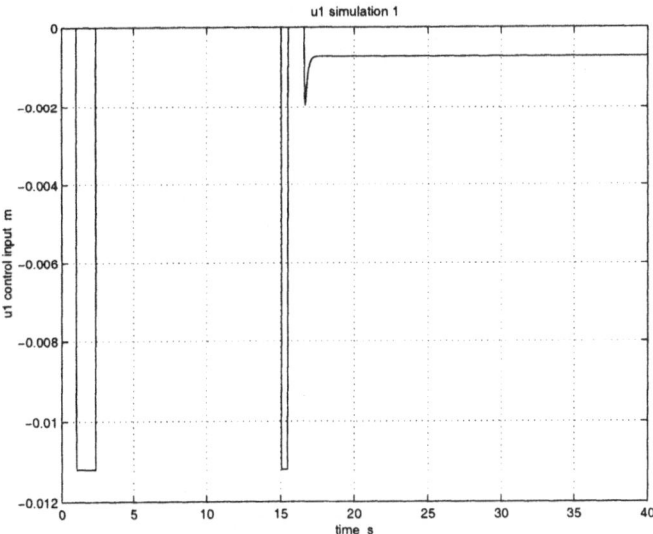

Figure 13.14: u1 Input variation for simulation 1

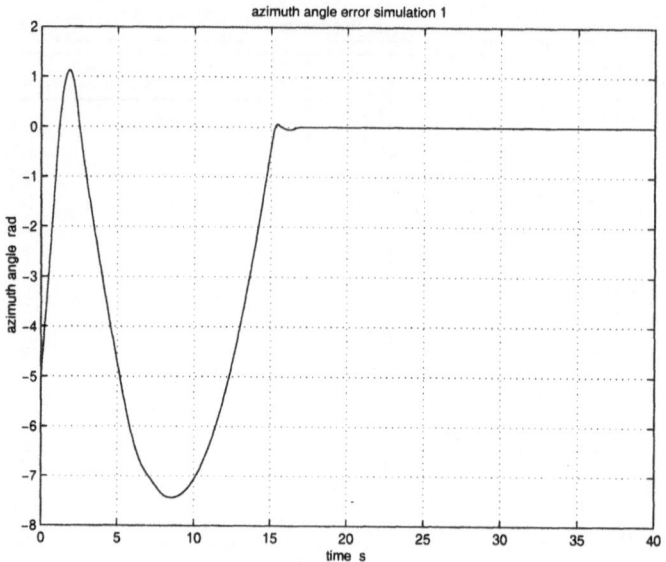

Figure 13.15: Azimuth angle error convergence for simulation 1

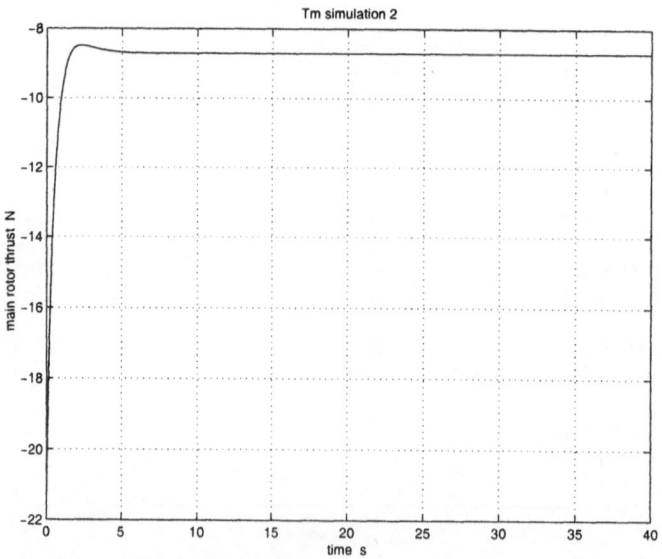

Figure 13.16: Thrust variation for simulation 2

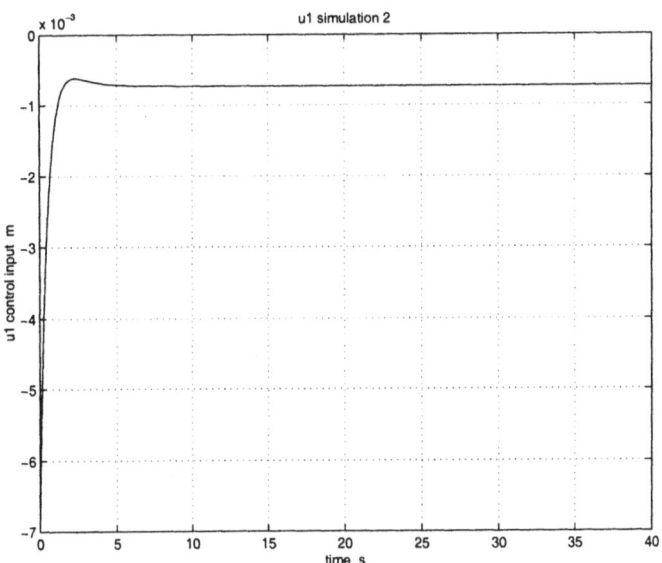

Figure 13.17: u1 Input variation for simulation 2

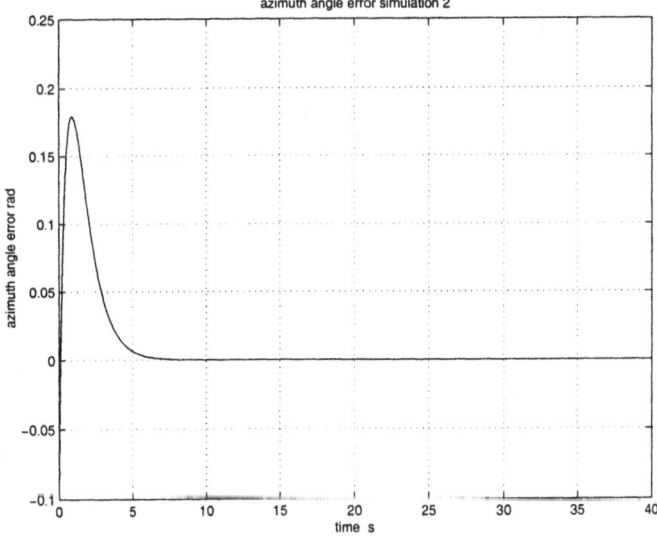

Figure 13.18: Azimuth angle error convergence for simulation 2

Chapter 14

Lagrangian helicopter model

14.1 Introduction

There is a growing interest in the construction and control of autonomous model helicopters [47, 90]*. Recently, a number of authors from the control community have begun to investigate an integrated non-linear dynamic model of a scale model autonomous helicopter (cf. conference papers [28, 62, 95, 101, 119] and more recently the journal papers [93, 102]). Model helicopters display a considerably different dynamic response than full scale helicopters. For example, the classical model [87, pg. 557] used for a full size helicopter does not model the interaction of the rotor blade dynamics with the rigid body dynamics of the airframe. Instead, the rotor blade dynamics are incorporated into the modelling of a daunting collection of aerodynamic and parasitic forces, which in turn act on the rigid body dynamics. It appears from experience that the regulation of the rotor speed of a model helicopter is an important part of the integrated control problem [118].

A simple dynamic representation of the full behaviour of a helicopter in all flight modes does not exist due to the different nature of the various aerodynamic forces in different flight conditions. As a consequence, a reasonable approach to the general control problem is to consider each

---

*The authors of this chapter are Robert Mahony and Rogelio Lozano. R. Mahony is with the Department of Electrical & Computer Systems Engineering, Monash University, Clayton, Victoria, 3800, Australia. R. Lozano is with the Laboratory Heudiasyc, UTC UMR CNRS 6599, Centre de Recherche de Royallieu, BP 20529, 60205 Compiègne Cedex, France.

flight condition separately, develop practical control laws for the various models based on foreseen mission requirements and then combine these into a practical control algorithm using hybrid control theory. For several reasons, the dynamics of a helicopter for manoeuvres close to hover are the simplest of all the possible cases to consider. Firstly, the parasitic aerodynamic force due to relative wind velocity is negligible. Furthermore, the rotor flapping angles (which are used to induce the rotational torque for pitch and roll control) are algebraic functions of the cyclic pitch angles (commanded by the pilot). This is not true in the presence of relative wind. Studying manoeuvres close to hover is important for mission objectives involving hovering, take-off and landing manoeuvres. The above discussion motivates the interest in studying the modelling and control of a model helicopter in the particular case where the aerodynamic effects are trivialised (hover type manoeuvres) and in which the rotor dynamics are fully considered.

In this chapter, we present two main results. Firstly, we present a simple dynamic model for an autonomous model helicopter for manoeuvres close to hover. The model is based on a Lagrangian expression for the energy of the system and varies from classical analysis in that we simplify the aerodynamic effects significantly but do not simplify the mechanical interaction of the main rotor blades with the full rigid body dynamics of the airframe. The model is a first step in a more detailed modelling of the dynamics of a helicopter with control as the objective rather than analysis. The model obtained is in block pure feedback form [49] and consequently a stabilizing (and local path tracking) control law may be designed based on backstepping techniques. However, the desired variables which we wish to track do not all enter at the first level of backstepping, nor do all the blocks of the system have the same dimension. As a consequence, we introduce two variations to standard backstepping procedures. Firstly, all controls are adjusted to have the same relative degree with respect to the first block of the block pure feedback system. This is done by dynamically extending those inputs that occur earlier in the cascade. Secondly, at each stage of backstepping, the full error is preserved (rather than assigning available controls directly and backstepping the remaining variables separately) and augmented by additional path tracking errors that are added when the appropriate relative degree requirements are satisfied. In this manner, a control law is designed that makes the overall system passive from the dynamically extended controls to a set of outputs derived from the tracking error. There is a considerable advantage associated with the proposed design procedure in comparison to the more classical backstepping approach, where each variable is separately considered, due to the saving in the algebraic complexity of the control design and the increased robustness associated with not having to cancel introduced non-linearities. By approaching the problem in this manner, it is possible to develop a unified control strategy that stabilizes the position and orientation of a helicopter as well as regulating the rotation of the main rotor blades. This is in contrast to many published control laws, where these tasks are separately considered [118].

Finally, we would like to point out that dynamic reduction type designs [31] should allow the results obtained to be extended directly to deal with secondary aerodynamic effects. For example, peaking analysis designs for simplified models of VTOL jump jets [91, pg. 246.] are based on a direct passivity controller for the unperturbed system followed by a peaking analysis. Analogously, we expect that by reducing the gains on the linear stability response in accordance with a peaking analysis and effective high gain type design (cf. [91, pg. 239]), the passivity-based controller designed in this chapter should be valid over a large range of initial conditions and desired paths.

The chapter consists of five sections. After the introduction, Section 14.2 presents the model considered in the sequel. In Section 14.3, a Lyapunov control law is derived based on a modified backstepping procedure. Section 14.4 presents an analysis of the proposed control, which provides some insight into the effect of the rotor dynamics on the overall helicopter dynamics. Two simulations are presented, showing the effects of unknown air resistance terms in the control design. The final section contains some brief conclusions.

14.2 Helicopter model

In this section, a Lagrangian model is derived for an autonomous model helicopter in terms of a local coordinate representation. The model considered is valid for moderate trajectory tracking, where the complex aerodynamic forces associated with the rotor response may be approximately compensated for by a static non-linear transformation of the control inputs.

Consider Figure 14.1. Denote the body or airframe of the helicopter by the letter **A**, the main rotor blades by the letter **B** and the tail rotor by the letter **C**. In addition, the helicopter as a whole is labelled by the letter **H**. Let $\mathcal{I} = \{E_x, E_y, E_z\}$ denote a right hand inertial frame stationary with respect to the earth and such that E_z denotes the vertical direction (cf. Figure 14.1). Let the vector $\xi = (x, y, z)$ denote the position of the center of mass of the helicopter relative to the frame $\mathcal{I}$ and a fixed (but arbitrary) zero point lying on the surface of the earth.

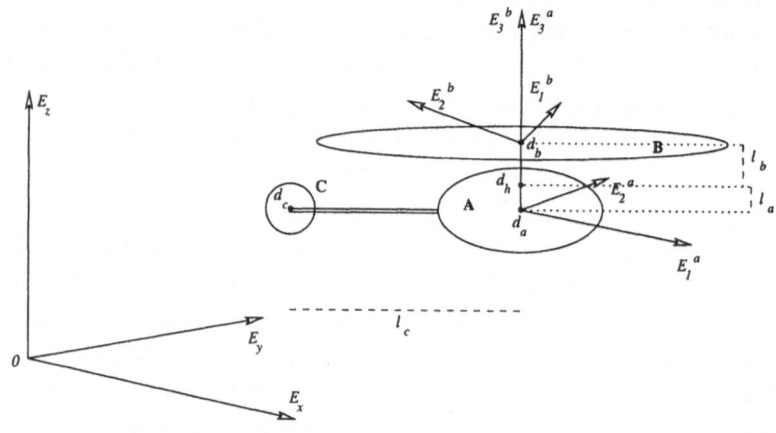

Figure 14.1: Diagram showing some of the notation used in the model of the Helicopter

Let $\mathcal{A} = \{E_1^a, E_2^a, E_3^a\}$ be a (right hand) body fixed frame for **A**. We choose E_1^a to correspond to the normal direction of flight of the helicopter, E_2^a is orthogonal and in the horizontal plane, while E_3^a should (hopefully) correspond with E_z in normal stationary hover conditions (cf. Figure 14.1). Analogous to the above, define $\mathcal{B} = \{E_1^b, E_2^b, E_3^b\}$ to be a body fixed frame for **B**.

Force control is obtained from lift due to the main rotor blades. The direction of the actual lift is oriented perpendicular to the orientation of the main rotor disk. In a full sized helicopter, the flexibility of the rotor blades is such that the orientation of the main rotor disk is not fixed perpendicular to the hub axis of the rotor blades. In fact, the aerodynamic and centrifugal forces acting on the rotor blades are more than 100 times stronger than the forces associated with the rigidity of the rotor blades and consequently, given a non-zero cyclic pitch input, the main rotor disk will quickly deform (in around the time it takes for a single rotation of the rotor blades, usually less than a second for a full sized helicopter) to balance the aerodynamic and centrifugal forces. This effect still results in torque control over pitch and roll directions since the tilt of the main rotor thrust itself generates rotational torque around the center of mass of the helicopter. The rotational torque obtained is directly proportional to the component of the main rotor thrust orthogonal to the principal lift direction $-E_3^a$ and the offset between the main rotor hub and the center of mass.

For a model helicopter, the situation is somewhat more complicated due to the relative rigidity of the rotor blades in comparison to their size and the centrifugal and aerodynamic forces that act on them. There is clearly a non-trivial component of the torque force generated by a cyclic pitch input that is derived directly from the mechanical coupling of the rotor blades to the rotor hub and centrifugal forces. Along with the low inertia of a model helicopter (and correspondingly small torque inputs required to obtain desired rotation), the rotor blade deflection due to cyclic pitch input is small in comparison to that obtained for a full sized helicopter. Even for a full sized helicopter, the measured rotor deflection angles tend to be less than 5 to 7 degrees. Thus, to a first approximation, it is possible to assume that the rotor blades are fixed rigidly to the rotor hub without flexibility and that the corresponding torques are generated directly from aerodynamic forces acting on the rigid rotor blades. Though this assumption is clearly an approximation, it should be accurate to within the precision of the control tasks considered. The actual control inputs applied to a physical model helicopter must be computed with respect to the full analysis of the rotor blade flexibility. As long as the manoeuvres considered are not too aggressive, then the relative wind effects are negligible and the secondary aerodynamic effects may be ignored. In this case, the transformation between a desired torque input to the full mechanical system and the control inputs is a static non-linear relationship. In this chapter, we consider the physical torques directly as inputs and assume that the transformation between the desired inputs specified by the control design and the actual physical control signals is a known algebraic function.

The discussion given above is summarised in the following formal assumptions

Assumption 14.1   *i) Due to the high angular velocity of the main rotor blades, they are modelled as a disk rather than as separate rotating blades. This disk is termed the main rotor disk and is assumed to rotate in an anti-clockwise direction when viewed from above.*

    ii) Since only moderate manoeuvres are considered, the flexibility of the rotor blades is ignored and the rotor disk is modelled as an infinitely stiff disk with infinitesimal thickness. The rotor disk is assumed to always rotate in the plane perpendicular to E_3^a. We assume that $E_3^b = E_3^a$.

iii) The tail rotor is considered to have no angular inertia. The aerodynamic force exerted by the tail rotor on the helicopter airframe is assumed to be purely a rotational torque around the axis E_3^a.

iv) The lift provides a translation force F_ξ (heave) acting directly through the center of mass of the airframe and permanently oriented in the direction E_3^a (cf. Figure 14.1).

v) Orientation control is modelled as three independent torques $\{\Gamma_1, \Gamma_2, \Gamma_3\}$ around each of the three body fixed frame directions $\{E_1^a, E_2^a, E_3^a\}$ (cf. Figure 14.1). These torques are applied directly to the airframe and do not result in any translational forces (small body forces) associated with secondary aerodynamic effects or flexibility of the rotor blades.

vi) The magnitude of all torques and forces generated by the rotor blades are modelled directly by control inputs. The actual control inputs for a physical system are generated via an algebraic relationship that depends only on known data, including the full state of the system and perhaps some knowledge of conditions.

vii) It is assumed that the engine dynamics operate on a time scale much faster than the airframe dynamics, or that they are insignificant with respect to the airframe dynamics, or that a sufficiently good model of the engine dynamics is known, which can be inverted to allow the engine torque to be directly applied to the blades.

viii) The only air resistance modelled is a simple drag force acting to slow the main rotor blades of the helicopter.

ix) The earth is assumed to be flat.[1]

■

Due to Assumption 14.1(iii), the dynamics of the tail rotor need not be separately modelled. The tail rotor is thought of purely as a means to generate a torque around the axis E_3^a and to counteract the effect of the drag on the main rotor blades.

---

[1]Navigation is the least of our worries.

The orientation of the helicopter airframe is given by three Euler angles

$$\nu = (\phi, \theta, \psi) \tag{14.1}$$

which are the classical "yaw", "pitch" and "roll" Euler angles commonly used in aerodynamic applications [33, pg. 608]. Firstly, a rotation of angle ϕ around the axes E_z is applied, corresponding to "yaw". Secondly, a rotation of angle θ around the rotated version of the E_y axis is applied, corresponding to "pitch" of the airframe. Lastly, a rotation of angle ψ around the axes E_1^a is applied. This corresponds to "roll" of $\mathbf{A}$ around the natural axis E_1^a. It should be noted that the Euler angles $\nu = (\phi, \theta, \psi)$ are not a global coordinate patch on $SO(3)$. Indeed, once $\theta \geq \frac{\pi}{2}$, then the correspondence between the Euler coordinates and the rotation matrices in $SO(3)$ is no longer one-to-one. Using Euler angles to represent the system dynamics will not be a problem for moderate manoeuvres.

The rotation matrix $R(\phi, \theta, \psi) \in SO(3)$ representing the orientation of the airframe $\mathbf{A}$ relative to a fixed inertial frame is[2]

$$R := R(\phi, \theta, \psi) = \begin{pmatrix} c_\theta c_\phi & s_\psi s_\theta c_\phi - c_\psi s_\phi & c_\psi s_\theta c_\phi + s_\psi s_\phi \\ c_\theta s_\phi & s_\psi s_\theta s_\phi + c_\psi c_\phi & c_\psi s_\theta s_\phi - s_\psi c_\phi \\ -s_\theta & s_\psi c_\theta & c_\psi c_\theta \end{pmatrix} \tag{14.2}$$

In addition, the relative angle of the rotor blades $\mathbf{B}$ to the helicopter airframe $\mathbf{A}$ is required to add the dynamics of the rotors. This angle is denoted as γ and is given by

$$\gamma = \cos^{-1}\langle E_1^b, E_1^a \rangle = \cos^{-1}\left((E_1^b)^T E_1^a \right)$$

where E_1^b is an axis of a body fixed frame for $\mathbf{B}$ lying in the plane of the rotor disk.

The generalized coordinates for the helicopter $\mathbf{H}$ are

$$q = (x, y, z, \phi, \theta, \psi, \gamma) \in \mathbb{R}^7$$

The generalized coordinates partition naturally into translational and rotational coordinates

$$\xi = (x, y, z) \in \mathbb{R}^3, \quad \eta = (\phi, \theta, \psi, \gamma) \in \mathbb{R}^4 \tag{14.3}$$

---

[2]The following shorthand notation for trigonometric function is used:

$$c_\beta := \cos(\beta), \quad s_\beta := \sin(\beta)$$

Remark 14.1 *If the tilt of the main rotor disk was considered explicitly, then it would be necessary to add two extra coordinates to represent its orientation with respect to the helicopter airframe and deal with the additional complications arising from the misalignment of the main rotor disk in deriving the expression for the energy discussed below. Due to the extreme time scale separation between the dynamics associated with the flapping angles (tilt of the main rotor disk) and the rigid body dynamics of the helicopter, it is not clear that a full Lyapunov model would have much practical validity.* ∎

The translational kinetic energy of the helicopter is

$$T_{\text{trans}} := \frac{m}{2} \langle \dot{\xi}, \dot{\xi} \rangle$$

where m denotes the mass of **H**.

Let $\mathbf{I}_A$ denote the inertia of the airframe **A**, with respect to the center of mass of **H**, expressed (as a constant matrix) in the body fixed frame $\mathcal{A}$. Let Ω_a denote the angular velocity of the airframe in the body fixed frame $\mathcal{A}$. Then the rotational kinetic energy of **A** is

$$T_{\text{rot}}^{\mathbf{A}} := \frac{1}{2} \langle \Omega_a, \mathbf{I}_A \Omega_a \rangle$$

Due to Assumption 14.1(ii), the orientation of $\mathcal{B}$ is obtained directly from the orientation of $\mathcal{A}$ by an additional rotation of angle γ around the E_3^a axis. Thus, the angular velocity of **B** expressed in the frame $\mathcal{A}$ is

$$\Omega_b = \Omega_a + \dot{\gamma} e_3$$

Here, e_3 is the unit vector with a one in the third position. In the frame $\mathcal{A}$, then $e_3 = E_3^a$, however, the above notation emphasises that the expression obtained is algebraic.

The inertia of the rotor blades in the frame $\mathcal{A}$ may be approximated by that of a disk rotating about a point not at its center of mass. However, since the geometry of a helicopter ensures that the center of mass of the airframe **A** and the blades **B** both lie on the axis E_3^a, then the center of mass of the helicopter **H** must also be co-linear with these points. Thus, the inertia matrix of the rotor disk, relative to the center of mass of **H**, and expressed in the body fixed frame $\mathcal{A}$ is a constant diagonal matrix

$$\mathbf{I}_B := \text{diag}(I_1^b, I_2^b, I_3^b)$$

Here, I_3^b is the moment of inertia of the blades around their main axis of rotation. The combined rotational kinetic energy of **A** and **B** can now be written directly as

$$
\begin{aligned}
T_{\text{rot}} &:= \frac{1}{2}\langle \Omega_a, \mathbf{I}_A \Omega_a \rangle + \frac{1}{2}\langle \Omega_b, \mathbf{I}_B \Omega_b \rangle \\
&= \frac{1}{2}\langle \Omega_a, \mathbf{I}_A \Omega_a \rangle + \frac{1}{2}\langle (\Omega_a + \dot{\gamma} e_3), \mathbf{I}_B(\Omega_a + \dot{\gamma} e_3) \rangle \\
&= \frac{1}{2}\langle \Omega_a, (\mathbf{I}_A + \mathbf{I}_B)\Omega_a \rangle + \frac{1}{2}\dot{\gamma}^2 I_3^b + \dot{\gamma}\langle \Omega_a, \mathbf{I}_B e_3 \rangle
\end{aligned}
$$

Let

$$
\mathbf{I}_H := \begin{pmatrix} & & & \vdots & 0 \\ & \mathbf{I}_A + \mathbf{I}_B & & \vdots & 0 \\ & & & \vdots & I_3^b \\ \cdots & \cdots & \cdots & \cdot & \cdot \\ 0 & 0 & I_3^b & \cdot & I_3^b \end{pmatrix} \in \mathbb{R}^{4\times4} \qquad (14.4)
$$

Due to its construction, the matrix $\mathbf{I}_H$ is positive definite. Furthermore, it is easily verified that

$$
T_{\text{rot}} = \frac{1}{2}\langle (\Omega_a, \dot{\gamma}), \mathbf{I}_H(\Omega_a, \dot{\gamma}) \rangle
$$

An angular velocity in the body fixed frame $\mathcal{A}$ is related to the generalized velocities $(\dot{\phi}, \dot{\theta}, \dot{\psi})$ (in the region where the Euler angles are valid) via the standard kinematic relationship [33, pg. 609]

$$
\Omega_a = \begin{pmatrix} \dot{\psi} - \dot{\phi}s_\theta \\ \dot{\theta}c_\psi + \dot{\phi}c_\theta s_\psi \\ \dot{\phi}c_\theta c_\psi - \dot{\theta}s_\psi \end{pmatrix}
$$

Defining

$$
W_\nu := \begin{pmatrix} -s_\theta & 0 & 1 & 0 \\ c_\theta s_\psi & c_\psi & 0 & 0 \\ c_\theta c_\psi & -s_\psi & 0 & 0 \\ 0 & 0 & 0 & 1 \end{pmatrix} \qquad (14.5)
$$

then

$$
\dot{\eta} = \begin{pmatrix} \dot{\phi} \\ \dot{\theta} \\ \dot{\psi} \\ \dot{\gamma} \end{pmatrix} = W_\nu^{-1}\begin{pmatrix} \Omega_a \\ \dot{\gamma} \end{pmatrix}
$$

Remark 14.2 *Note that*

$$\det(W_\nu) = -\cos(\theta)$$

and thus the kinematic transformation between angular velocity Ω and its representation in generalized coordinates $\dot\eta$ is non-singular for all orientations except those where $\theta = \frac{\pi}{2}$.   ∎

Define

$$\mathbb{I} := \mathbb{I}(\eta) = W_\nu^T \mathbf{I}_H W_\nu \tag{14.6}$$

and observe that

$$T_{\text{rot}} := \frac{1}{2}\langle W_\nu \dot\eta, \mathbf{I}_H W_\nu \dot\eta \rangle = \frac{1}{2}\langle \dot\eta, \mathbb{I}\dot\eta \rangle$$

Thus, the matrix $\mathbb{I} := \mathbb{I}(\eta)$ acts as the inertia matrix for the full rotational kinetic energy of the helicopter expressed directly in terms of the generalized coordinates η.

The only potential that needs to be considered is the standard gravitational potential given by

$$U = mgz$$

where we recall that $\xi = (x, y, z)$ is the center of mass of the helicopter. The full Lagrangian function $\mathcal{L}$ is now

$$\begin{aligned}
\mathcal{L}(q, \dot q) &:= T_{\text{trans}} + T_{\text{rot}} - U \\
&= \frac{1}{2}m\dot\xi^T\dot\xi + \frac{1}{2}\dot\eta^T\mathbb{I}\dot\eta - mgz
\end{aligned} \tag{14.7}$$

External forces and torques applied to the helicopter airframe are due to aerodynamic lift generated by the rotor blades. The angle of attack of the rotor blades is varied systematically during rotation according to the orientation of the swash plate. The lift generated can be manipulated to yield an overall lift (termed heave), and two differential torques that result in pitch and roll of the airframe. A third independent torque is provided by the tail rotor.

According to Assumption 14.1(iv), there is a single translational force acting in direction E_3^a. Thus, the direction of the applied translational force for **A** is determined by the orientation of the airframe. Expressed in the inertial frame $\mathcal{I} = \{E_x, E_y, E_z\}$, the thrust direction is given by the vector

$$G(\eta) := Re_3 = \begin{pmatrix} c_\psi s_\theta c_\phi + s_\psi s_\phi \\ c_\psi s_\theta s_\phi - s_\psi c_\phi \\ c_\psi c_\theta \end{pmatrix} \in \mathbb{R}^3 \qquad (14.8)$$

where e_3 is the unit vector with a one in the third position and zeros elsewhere and R is given by (14.2). Note that $|G(\eta)| = 1$ since it is the rotation of a unit vector. According to Assumption 14.1(vi), the magnitude of the applied translational force may be modelled directly by a control input $u \in \mathbb{R}$. Thus, the applied generalized force on the coordinates ξ may be written

$$F_\xi := uG(\eta). \qquad (14.9)$$

Torques applied to the helicopter **H** translate into generalized forces on the coordinates (ϕ, θ, ψ). The torques $\{\Gamma_1, \Gamma_2, \Gamma_3\}$ are applied around the axis E_1^a, E_2^a and E_3^a, which are unit base vectors in the frame $\mathcal{A}$ (cf. Figure 14.1). According to Assumption 14.1(vi), torques are represented directly as linear control inputs. The motor of the helicopter applies a torque directly to the rotor blades **B**, which once again is modelled as a linear input Γ_4. This in turn generates a reactionary torque, acting on the airframe around the axis E_3^a, of equal magnitude and opposite sign. Thus, the torques applied to the rigid body motion of the airframe **A**, expressed in the frame $\mathcal{A}$, are

$$\begin{pmatrix} \Gamma_1 \\ \Gamma_2 \\ \Gamma_3 - \Gamma_4 \end{pmatrix}$$

while the force applied to rotate the blades, and thus appearing as a generalized force acting on γ, is $+\Gamma_4$. In addition to the generalized force $+\Gamma_4$ applied to the blades, there is a drag term $-\Sigma$ (which in equilibrium conditions exactly cancels Γ_4) due to the air resistance of the rotor blades. To convert these torques into generalized forces on the coordinates $(\phi, \theta, \psi, \gamma)$, it is necessary to apply the inverse kinematic relationship W_ν^{-1} to map velocities and torques expressed in the frame $\mathcal{A}$ into generalized velocities and forces in the coordinates q. Thus, the generalized forces on the η variables are

$$\tau^0 := \begin{pmatrix} \tau_\phi \\ \tau_\theta \\ \tau_\psi \\ \tau_\gamma \end{pmatrix} = W_\nu^{-1} \begin{pmatrix} \Gamma_1 \\ \Gamma_2 \\ \Gamma_3 - \Gamma_4 \\ \Gamma_4 - \Sigma \end{pmatrix} \qquad (14.10)$$

Assuming that a good model of the drag Σ is known, then this term may be cancelled using Γ_4 and then Γ_3 chosen to cancel the components of Γ_3 in the "yaw" control. Thus, the ideal torque input, denoted τ^0, may be thought of as a free input. In reality, the cancellation of Σ will be imprecise and there will always be an associated modelling error. Let Σ_0 denote the best existing model of the air resistance torque and let σ_t denote the residual time-signal. Thus

$$\Sigma := \Sigma_0 + \sigma_t$$

We design a control for the system based on the information Σ_0 and then look at the effect of the unmodelled term σ_t considered as a disturbance to the closed-loop system. Since the control design is based on an energy perspective, it is expected that the performance of the overall controller will still be good for moderate disturbances σ_t and this is observed in the simulations presented in Section 14.4. Assuming that the orientation of the airframe remains in the region where the Euler coordinates are well defined, then the relationship given by (14.10) is one-to-one and one can work directly with the inputs $\tau \approx (\tau_\phi, \tau_\theta, \tau_\psi, \tau_\gamma)$ with the understanding that there will be a slight perturbation present in the actual closed-loop system

$$\tau = \tau^0 - \sigma_t W_\nu^{-1} e_4 = \tau^0 - \sigma_t e_4 \tag{14.11}$$

The model for the full helicopter dynamics is obtained from the Euler-Lagrange equations with external generalized forces

$$\frac{d}{dt}\frac{\partial \mathcal{L}}{\partial \dot{q}} - \frac{\partial \mathcal{L}}{\partial q} = F \tag{14.12}$$

where $F = (F_\xi, \tau)$. Since the Lagrangian function $\mathcal{L}$ contains no cross terms in the kinetic energy combining $\dot{\xi}$ with $\dot{\eta}$, the Euler Lagrange equation (14.12) can be partitioned into dynamics for the ξ coordinates and the η dynamics. One obtains

$$m\ddot{\xi} + mge_3 = F_\xi = uG(\eta) \tag{14.13}$$

$$\mathbb{I}\ddot{\eta} + C(\eta, \dot{\eta})\dot{\eta} = \tau \tag{14.14}$$

where the term $C(\eta, \dot{\eta})\dot{\eta}$ is referred to as the Coriolis terms and contains the gyroscopic and centrifugal terms associated with the η dependence of $\mathbb{I}$ and can be computed using the classical equations (cf. for example [110, Equation (6.3.12), pg. 142]). An explicit form for the Coriolis matrix is given in Section 14.4. If the perturbation term (14.11) is included, then one obtains

$$\mathbb{I}\ddot{\eta} + C(\eta, \dot{\eta})\dot{\eta} = \tau = \tau^0 - \sigma_t e_4 \qquad (14.15)$$

It appears that the perturbation σ_t enters only into the dynamics of the fourth component γ, however, due to the coupling in the inertia matrix $\mathbb{I}$, it also directly affects the dynamics of the yaw ϕ. For a diagonal inertia matrix $\mathbf{I}_A = \text{diag}(I_1^a, I_2^a, I_3^a)$, then the influence of this term can be seen by computing the fourth column of the inverse inertia matrix $\mathbb{I}^{-1}$

$$\mathbb{I}_{(\cdot,4)}^{-1} = \frac{1}{c_\theta^2 I_3^a I_3^b} \begin{pmatrix} -c_\theta c_\psi I_3^b \\ c_\theta^2 s_\psi I_3^b \\ -s_\theta c_\theta c_\theta I_3^b \\ c_\theta^2 (I_3^a + I_3^b) \end{pmatrix} \qquad (14.16)$$

This coupling is much stronger in the case of a model helicopter than in the case of a full sized helicopter due to the relative importance of the inertia I_3^b of the blades compared to the inertia I_3^a of the airframe around the E_3^a axis. This is the reason why many of the standard control schemes developed for full sized helicopters (which use a separate control loop to regulate the rotor dynamics) have not performed satisfactorily for model helicopter applications.

14.3 Energy-based control design

In this section, a control law is proposed for the helicopter model introduced in the previous section. The algorithm is based on the back-stepping methodology [49], though several novel modifications of the standard backstepping methodology are introduced to deal with the particular structure encountered.

The problem considered is that of smooth path tracking. In particular, we consider a given path in the coordinates $\xi = (x, y, z)$ and look for a control law that manipulates the full generalized coordinates to ensure that the path is followed. In addition to the path coordinates in ξ, we add additional trajectory requirements on the yaw angle $\phi(t)$ and on the regulation of the rotor speed $\dot{\gamma}$. The specified path is practically motivated by the desire to regulate the position, orientation and rotor speed of a helicopter in hover. The desired trajectories do not fit into the standard framework for backstepping path tracking designs.

Definition 14.1 Consider the model of a helicopter given by (14.13) and (14.14). Let

$$\hat{\xi} : \mathbb{R} \to \mathbb{R}^3$$
$$\hat{\phi} : \mathbb{R} \to \mathbb{R}$$

be smooth trajectories $\hat{\xi}(t) := (\hat{x}(t), \hat{y}(t), \hat{z}(t))$ and $\hat{\phi}(t)$. Let $\kappa > 0$ be a constant. The control problem considered is:

> Find a feedback control action $(u, \tau) \in \mathbb{R}^4$ depending only on the measurable states $(\dot{\xi}, \xi, \dot{\eta}, \eta)$ and arbitrarily many derivatives of the smooth trajectory $(\hat{\xi}(t), \hat{\phi}(t))$ such that the tracking error
>
> $$\mathcal{E} := (\xi(t) - \hat{\xi}(t), \phi(t) - \hat{\phi}(t), \dot{\gamma}(t) - \kappa) \in \mathbb{R}^5 \qquad (14.17)$$
>
> is asymptotically stable.

∎

Two points need to be emphasised regarding the control problem as stated. Firstly, the desired trajectory must be sufficiently smooth before the techniques employed in the sequel may be applied. We will be keeping in mind the problem of regulation to a set point for hover regulation as the prime example. Secondly, the tracking error as defined is a mixture of paths in the translation coordinates, the orientation coordinates and derivatives of the orientation coordinates. These errors all have different relative degrees with respect to the control inputs and preclude the direct application of standard backstepping techniques.

Consider once again (14.13) and (14.14). It is clear that these equations are in block pure feedback form [49, pg. 61], where the first block is (14.13). This leads one to consider a partial error in the variables $\xi = (x, y, z)$ and to use this to backstep, adding additional error variables as appropriate.

Consider the error

$$e := \xi(t) - \hat{\xi}(t) \qquad (14.18)$$

Then following the standard approach for path tracking in mechanical systems [105, pg. 398], we consider the output

$$\alpha := \dot{e} + e \qquad (14.19)$$

The choice of α is motivated by a zero dynamics argument. That is, if one designs a controller to drive $\alpha \to 0$, then the zero dynamics

$$\dot{\alpha} = -\alpha$$

are globally and asymptotically stable and ensure that the error itself will also converge to zero.

Taking the time derivative of α and substituting for (14.13) yields

$$m\frac{d}{dt}\alpha = m\ddot{e} + m\dot{e}$$

$$= m(\dot{e} - \ddot{\hat{\xi}}) + m\ddot{\xi}$$

$$= m(\dot{e} - \ddot{\hat{\xi}}) - mge_3 + G(\eta)u \qquad (14.20)$$

Formally, there is only a single input "u" present in this equation and it is impossible to assign the desired three-dimensional stable dynamics. The process of backstepping suggests that we consider the variables η as inputs themselves. In this case, it is clear that the unit vector $G(\eta)$ may be arbitrarily assigned direction and that the control u can be used to assign the magnitude desired for the stable dynamics required. Unfortunately, such an approach brings its own problems since formally there are now four input variables (η, u) to assign three-dimensional dynamics. Moreover, solving the vector $G(\eta)$ for the angles (ϕ, θ, ψ) introduces some unpleasant non-linearities if the resulting explicit expressions are used in a backstepping design. An indication of the complications involved in an approach like this are present in the design of explicit backstepping control of a VTOL aircraft [91, pg. 246 and references]. Rather than take this approach, we will view the vector $G(\eta)u$ as a vector in $\mathbb{R}^3$ and carry the full expression through to the backstepping procedure. Thus, we define an error

$$\beta_1 = G(\eta)u - X \qquad (14.21)$$

where

$$X := X(\ddot{\hat{\xi}}, \dot{\hat{\xi}}, \hat{\xi}, \dot{\xi}, \xi)$$

is a function of known signals and is chosen to assign stable dynamics to the error α. In particular, choose

$$X = -\left(m(\dot{e} - \ddot{\hat{\xi}}) - mge_3 + \alpha\right) \qquad (14.22)$$

Consider the storage function

$$S_\alpha = \frac{m}{2}|\alpha|^2 = \frac{m}{2}\alpha^T\alpha \qquad (14.23)$$

Differentiating the storage S_α, one obtains

$$\begin{aligned}
\dot{S}_\alpha &= -|\alpha|^2 + \alpha^T\left(G(\eta)u - X\right) \\
&= -|\alpha|^2 + \alpha^T\beta_1 \qquad (14.24)
\end{aligned}$$

In the formal process of backstepping the error, β_1 would now be differentiated and stable dynamics assigned to it in turn. Such an approach requires that time derivatives of the input u are computed. This can be achieved by dynamically extending the input u so that it has the same relative degree in (14.13) as the variables ν. Thus, we rewrite the helicopter dynamics, (14.13) and (14.14), and augment these dynamics with a cascade of two integrators feeding into the control action u

$$\begin{aligned}
m\ddot{\xi} &= G(\eta)u - mge_3 & (14.25) \\
\dot{\eta} &= \eta_2 & (14.26) \\
\dot{u} &= u_2 & (14.27) \\
\dot{\eta}_2 &= \ddot{\eta} = \mathbb{I}^{-1}\tau - \mathbb{I}^{-1}C(\eta,\dot{\eta})\dot{\eta} & (14.28) \\
\dot{u}_2 &= \ddot{u} = v & (14.29)
\end{aligned}$$

where the new variable $v \in \mathbb{R}$ along with the original generalized torques τ are the inputs for the augmented system. Note that the v and τ inputs now both have a relative degree of four with respect to the ξ coordinates. Moreover, note that the error α has a relative degree of three with respect to the inputs (τ, v). The kinematic equations (14.26) and (14.27) are included to display explicitly the block pure feedback form of the equations. This occurs in three cascaded blocks, the first, (14.25), the second, (14.26) and (14.27) and the final block, (14.28) and (14.29).

The design error β_1 is an error in the coordinates η and the control u. It is easily seen that β_1 has a relative degree of two with respect to the inputs (τ, v). Recalling the additional tracking errors given in Definition 14.1, note that $(\phi - \hat{\phi})$ also has a relative degree of two while $(\dot{\gamma} - \kappa)$ has a relative degree of one. Thus, before continuing with the formal backstepping procedure, it is possible to augment β_1 with an additional term that accounts for the tracking performance of the yaw $(\phi - \hat{\phi})$. The final tracking error $(\dot{\gamma} - \kappa)$ will be saved until the last step of the design procedure. Consider the augmented design error

$$\beta = (\beta_1^T, (\phi - \hat\phi))^T = \begin{pmatrix} G(\eta)u - X \\ \phi - \hat\phi \end{pmatrix} \in \mathbb{R}^4 \qquad (14.30)$$

The design error $\beta := \beta_t(\nu, u)$ can be thought of as a time-varying function of (ν, u). It is a straightforward, though somewhat tedious calculation, to show that in a suitable neighbourhood of $(\nu, u) = (\phi, 0, 0, u_0)$ for non-zero u_0 then the error β is a diffeomorphism. That is, it maps the four variables (ϕ, θ, ψ, u) locally one-to-one into $\mathbb{R}^4$. Thus, the addition of the extra tracking condition $(\phi \rightarrow \hat\phi)$ has removed the difficulty associated with uniquely defining a trajectory for the variables ν that arose when first defining β_1. Given that the map β is locally a diffeomorphism on the domain of interest, the Jacobian

$$J(\nu, u) := \frac{\partial\beta}{\partial(\nu, u)}(\nu, u) \in \mathbb{R}^{4\times 4}$$

is also well defined and non-singular in this domain.

Taking the time derivative of β yields

$$\frac{d}{dt}\beta = J(\nu, u) \begin{pmatrix} \dot\nu \\ \dot u \end{pmatrix} - \begin{pmatrix} \dot X \\ \dot{\hat\phi} \end{pmatrix} \qquad (14.31)$$

Note that

$$\dot X = -\dot\alpha - m(\ddot e - \hat\xi^{(3)}) = m\hat\xi^{(3)} - (1+m)\ddot e - \dot e$$
$$= m\hat\xi^{(3)} + (1+m)\ddot{\hat\xi} - \frac{(1+m)}{m}(G(\eta)u - mge_3) - \dot e$$
$$:= \dot X(\hat\xi^{(3)}, \dots, \hat\xi, \dot\xi, \xi, \eta, u)$$

Thus, one can define $Y = Y(\hat\xi^{(3)}, \dots, \hat\xi, \dot\xi, \xi, \eta, u)$ by

$$Y := \begin{pmatrix} \dot X \\ \dot{\hat\phi} \end{pmatrix} - \beta - \begin{pmatrix} \alpha \\ 0 \end{pmatrix} \in \mathbb{R}^4$$

Observe that

$$\beta^T \begin{pmatrix} \alpha \\ 0 \end{pmatrix} = \alpha^T\beta_1$$

since the fourth coordinate of β is cancelled by a zero. Consider the storage function

$$S_\beta := \frac{1}{2}|\beta|^2$$

then it is easily verified that

$$\dot{S}_\beta = -|\beta|^2 - \alpha^T \beta_1 + \beta^T \left(J(\nu, u) \begin{pmatrix} \dot{\nu} \\ \dot{u} \end{pmatrix} - Y \right)$$

Following the backstepping methodology, a third design error is defined

$$\delta_1 := \left(J(\nu, u) \begin{pmatrix} \dot{\nu} \\ \dot{u} \end{pmatrix} - Y \right) \tag{14.32}$$

The next step is to backstep once more with the error δ_1. However, before this is done, it is possible to augment the error δ_1 with an error to guarantee the final tracking requirement $\dot{\gamma} \to \kappa$. Once again, the tracking error for γ is introduced when its relative degree matches the relative degree of the design error propagated from an earlier block. Thus, let

$$\delta = \begin{pmatrix} \delta_1 \\ \dot{\gamma} - \kappa \end{pmatrix}$$

To simplify the following structure, define

$$\overline{J}(\eta, u) := \begin{pmatrix} J(\nu, u) & 0 \\ 0 & 1 \end{pmatrix}, \quad \overline{Y} := \begin{pmatrix} Y \\ \kappa \end{pmatrix} \tag{14.33}$$

then

$$\delta := \left(\overline{J}(\eta, u) S \begin{pmatrix} \dot{\eta} \\ \dot{u} \end{pmatrix} - \overline{Y} \right)$$

where $S \in \mathbb{R}^{5 \times 5}$ is a permutation matrix that exchanges entries four and five of $(\dot{\eta}, \dot{u})$, ensuring that the $\dot{\gamma}$ and $\dot{u}$ terms match up with the correct rows of $\overline{J}(\eta, u)$.

Consider the derivative of δ

$$\dot{\delta} = \overline{J}(\eta, u) S \begin{pmatrix} \ddot{\eta} \\ \ddot{u} \end{pmatrix} + \dot{\overline{J}}(\eta, u) S \begin{pmatrix} \dot{\eta} \\ \dot{u} \end{pmatrix} - \dot{\overline{Y}}$$

Though an explicit notation for the Jacobian $\overline{J}(\eta, u)$ is required so that the linear dependence of the second derivatives $(\ddot{\eta}, \ddot{u}, \ddot{\gamma})$ is clearly expressed, the remaining terms in the above expressions are left as simple time derivatives due to space restrictions. It should be noted, however, that the evaluation of $\dot{\overline{J}}(\eta, u)$ and $\dot{\overline{Y}}$ depends only on known signals.

Let $r > 0$ be a constant gain and define a vector $Z :=$ $Z(\hat{\xi}^{(4)}, \ldots, \dot{\xi}, \ddot{\xi}, \xi, \dot{\eta}, \eta, \dot{u}, u)$ by

$$Z = -\dot{\overline{J}}(\eta, u)S \begin{pmatrix} \dot{\eta} \\ \dot{u} \end{pmatrix} + \dot{\overline{Y}} - r\delta - \begin{pmatrix} \beta \\ 0 \end{pmatrix} \in \mathbb{R}^5$$

The gain $r > 0$ is introduced to assign a rate of convergence to the δ error dynamics that are independent of the rates of convergence chosen for the other error variables. The reason for this choice is related to a robustness analysis and is discussed in Section 14.4. Consider a third storage function

$$S_\delta = \frac{1}{2}|\delta|^2 \qquad (14.34)$$

Computing the time derivative of S_δ yields

$$\frac{d}{dt}S_\delta = -r|\delta|^2 - \beta^T \delta_1 + \delta^T \left(\overline{J}(\eta, u)S \begin{pmatrix} \ddot{\eta} \\ \ddot{u} \end{pmatrix} - Z \right)$$

Define

$$K(\eta, u) := \overline{J}(\eta, u)S \begin{pmatrix} \mathbb{I}^{-1} & 0 \\ 0 & 1 \end{pmatrix} \in \mathbb{R}^{5 \times 5} \qquad (14.35)$$

It follows directly from the invertibility of $J(\eta, u)$, S and $\mathbb{I}$ that $K(\eta, u)$ is invertible. Thus, using (14.28) and (14.29), one can write

$$\overline{J}(\eta, u)S \begin{pmatrix} \ddot{\eta} \\ \ddot{u} \end{pmatrix} = K(\eta, u) \begin{pmatrix} \tau \\ v \end{pmatrix} + \overline{J}(\eta, u)S \begin{pmatrix} -\mathbb{I}^{-1}C(\eta, \dot{\eta})\dot{\eta} \\ 0 \end{pmatrix}$$

Since the matrix $K(\eta, u)$ is full rank and the vector (τ, v) is a full rank vector of inputs, then the non-linear contribution to the δ dynamics may be explicitly cancelled by choosing

$$\begin{pmatrix} \tau \\ v \end{pmatrix} := K(\eta, u)^{-1}\overline{J}(\eta, u)S \begin{pmatrix} \mathbb{I}^{-1}C(\eta, \dot{\eta})\dot{\eta} \\ 0 \end{pmatrix} + K(\eta, u)^{-1}Z \quad (14.36)$$

As a consequence of applying this control, one has

$$\dot{S}_\delta = -r|\delta|^2 - \beta^T \delta_1$$

Proposition 14.1 *Consider the augmented dynamics of a helicopter given by (14.25)-(14.29) and assume a desired trajectory $(\hat{\xi}, \hat{\phi}, \kappa)$ is given according to Definition 14.1. Then, if the closed-loop trajectory evolves such that Euler angle representation of the airframe orientation remains well defined for all time t and the applied thrust u is never zero, then the control law (τ, v) given by (14.36) ensures exponential stabilization of the tracking error $\mathcal{E}$ in (14.17).* ∎

Proof 14.1 *Consider the combined storage function*

$$S = \frac{1}{2}(|e|^2 + S_\alpha + S_\beta + S_\delta)$$

Recalling the definition of α, one has

$$\dot{e} = -e + \alpha$$

Thus

$$\frac{1}{2}\frac{d}{dt}|e|^2 = -|e|^2 + e^T\alpha \tag{14.37}$$

$$= -\frac{1}{2}|e|^2 - \frac{1}{2}|e - \alpha|^2 + \frac{1}{2}|\alpha|^2 \tag{14.38}$$

The time derivative of S is computed by substituting the above calculation along with the expressions obtained earlier for the derivatives of S_α, S_β and S_δ.

$$\dot{S} = -\frac{1}{2}|e|^2 - \frac{1}{2}|e - \alpha|^2 + \frac{1}{2}|\alpha|^2 - |\alpha|^2 + \alpha^T\beta_1 - |\beta|^2 - \alpha^T\beta_1$$
$$+ \beta^T\delta_1 - r|\delta|^2 - \beta^T\delta_1$$
$$= -\frac{1}{2}|e|^2 - \frac{1}{2}|e - \alpha|^2 - \frac{1}{2}|\alpha|^2 - |\beta|^2 - r|\delta|^2$$

It follows that $\dot{S} < 0$ unless $e = 0 = \alpha = \beta = \delta$. Thus, by applying Lypunov's theorem, all the errors e, α, β and δ are asymptotically stable to zero and the desired tracking is achieved. ∎

14.4 Analysis and simulations

In this section, we derive the explicit equations for the helicopter based on the development presented in Section 14.2 and present some simple simulations indicating the performance of the proposed control algorithm.

A standard assumption in modelling full sized helicopters is that the principal axis of the inertia is aligned with the body fixed frame $\mathcal{A}$ [87, pg. 557]. This is a consequence of basic design principles based on ease of flying for a human pilot as well as ease of manufacture and is equally valid for model helicopters that are constructed to resemble full sized helicopters. Since the inertia associated with the main rotor disk (considered as a rigid object attached to the airframe) is also a diagonal matrix, the inertia matrix $\mathbf{I}_H$ may be written as

$$
\mathbf{I}_H := \begin{pmatrix} (I_1^a + I_1^b) & 0 & 0 & 0 \\ 0 & I_2^a + I_2^b & 0 & 0 \\ 0 & 0 & I_3^a + I_3^b & I_3^b \\ 0 & 0 & I_3^b & I_3^b \end{pmatrix}
$$

$$
= \begin{pmatrix} \lambda_1 & 0 & 0 & 0 \\ 0 & \lambda_1 - \lambda_2 & 0 & 0 \\ 0 & 0 & \lambda_1 - \lambda_2 - \lambda_3 & \lambda_4 \\ 0 & 0 & \lambda_4 & \lambda_4 \end{pmatrix}
$$

where $\lambda_1 = I_1^a + I_1^b$, $\lambda_2 = \lambda_1 - I_2^a - I_2^b$, $\lambda_3 = \lambda_1 - \lambda_2 - I_3^a - I_B^3$, $\lambda_4 = I_3^b$.
The constants λ_i are chosen to represent the relative ellipticity of the inertia of $\mathbf{H}$. If the inertia of the rigid object $\mathbf{H}$ corresponds to that of a sphere, then both the λ_2 and λ_3 terms are zero. Similarly, if just $\lambda_3 = 0$, then it indicates that the inertia around E_2^a and E_3^a is equal. From (14.5) and (14.6), it is a simple matter to compute the inertia matrix I_{rot}^h

$$
\mathbb{I}(\eta) := W_\nu^T I^h W_\nu
$$

$$
= \begin{pmatrix} I_{11} & \lambda_3 c_\theta s_\psi c_\psi & -\lambda_1 s_\theta & \lambda_4 c_\theta c_\psi \\ \lambda_3 c_\theta s_\psi c_\psi & I_{22} & 0 & -\lambda_4 s_\psi \\ -\lambda_1 s_\theta & 0 & \lambda_1 & 0 \\ \lambda_4 c_\theta c_\psi & -\lambda_4 s_\psi & 0 & \lambda_4 \end{pmatrix} \quad (14.39)
$$

where $I_{11} = \lambda_1 + \lambda_2 c_\theta^2 + \lambda_3 c_\theta^2 c_\psi^2$ and $I_{22} = \lambda_1 - \lambda_2 + \lambda_3 s_\psi^2$.
It is of interest to look at the particular form of the Coriolis matrix $C(\eta, \dot{\eta})$ that accounts for all gyroscopic and centrifugal forces for the helicopter. Following [110, pg. 142], one obtains

$$
C(\eta, \dot{\eta}) = \frac{\lambda_1 c_\theta}{2} \begin{pmatrix} 0 & -\dot{\psi} & -\dot{\theta} & 0 \\ \dot{\psi} & 0 & \dot{\phi} & 0 \\ -\dot{\theta} & -\dot{\phi} & 0 & 0 \\ 0 & 0 & 0 & 0 \end{pmatrix} + \frac{\lambda_2}{2} \begin{pmatrix} -\dot{\theta} s_{2\psi} & -\dot{\phi} s_{2\theta} & 0 & 0 \\ \dot{\phi} s_{2\theta} & 0 & 0 & 0 \\ 0 & 0 & 0 & 0 \\ 0 & 0 & 0 & 0 \end{pmatrix}
$$

$$
+ \frac{\lambda_3}{2} \begin{pmatrix} C_{311} & C_{312} & C_{313} & 0 \\ C_{321} & C_{322} & \dot{\phi} c_\theta c_{2\psi} & 0 \\ C_{331} & -\dot{\phi} c_\theta c_{2\psi} & 0 & 0 \\ 0 & 0 & 0 & 0 \end{pmatrix}
$$

$$
+ \frac{\lambda_4}{2} \begin{pmatrix} 0 & -\dot{\gamma} s_\theta c_\psi & -\dot{\gamma} c_\theta s_\psi & C_{414} \\ \dot{\gamma} s_\theta c_\psi & 0 & -\dot{\gamma} c_\psi & C_{424} \\ \dot{\gamma} c_\theta s_\psi & \dot{\gamma} c_\psi & 0 & C_{434} \\ C_{441} & C_{442} & C_{443} & 0 \end{pmatrix} \quad (14.40)
$$

where

$$C_{311} = -\left(\dot{\theta}s_{2\psi}c_\psi^2 + \dot{\psi}c_\theta^2 s_{2\psi}\right)$$

$$C_{312} = -\left(\dot{\phi}s_{2\theta}c_\psi^2 + \dot{\theta}s_{2\psi}s_\theta + \dot{\psi}c_\theta^2 s_{2\psi}\right)$$

$$C_{313} = -C_{331} = -\left(\dot{\phi}c_\theta^2 s_{2\psi} - \dot{\theta}c_\theta c_{2\psi}\right)$$

$$C_{321} = \left(\dot{\phi}s_{2\theta}c_\psi^2 + \dot{\psi}c_\theta c_{2\psi}\right)$$

$$C_{322} = \left(\dot{\phi}s_{2\psi}s_\theta + \dot{\psi}s_{2\psi}\right)$$

$$C_{414} = C_{441} = -(\dot{\theta}s_\theta c_\psi + \dot{\psi}c_\theta s_\psi)$$

$$C_{424} = (\dot{\phi}s_\theta c_\psi - \dot{\psi}c_\psi)$$

$$C_{434} = -C_{443} = -(\dot{\phi}c_\theta s_\psi + \dot{\theta}c_\psi)$$

$$C_{442} = (-\dot{\phi}s_\theta c_\psi - \dot{\psi}c_\psi)$$

For simplicity of notation, one may write

$$C(\eta, \dot{\eta}) = \lambda_1 C_1(\eta, \dot{\eta}) + \lambda_2 C_2(\eta, \dot{\eta}) + \lambda_3 C_3(\eta, \dot{\eta}) + \lambda_4 C_4(\eta, \dot{\eta})$$

where the matrices $C_1, \ldots, C_4$ are defined by reference to (14.40). It is easily verified that the matrix C_1 corresponds to the Coriolis matrix for a spherical rigid body in Euler coordinates. The contributions of C_2 and C_3 add the corrections for the ellipsoidal nature of the helicopter inertia, while C_4 provides the contributions to the overall inertia due to the rotation of the rotor blades. It is important to note that in C_4, the angular velocity of rotation of the blades enters into that part of the matrix that contributes dynamics to the rigid body rotation of the helicopter. Thus, it is to be expected that changes in the angular velocity of the rotor blades will effect the rigid body dynamics of the helicopter.

The complexity of the above expressions indicate the difficulties that will be encountered if one attempts to model the dynamics of a helicopter via a black box type approach. Much of the complexity is a consequence of the highly non-linear nature of the Euler coordinates for rigid body rotations. However, this non-linearity is an inherent part of the system and cannot be ignored. If one attempts to identify a linearized model of the helicopter in the vicinity of a stationary point, then the validity of the approximation will certainly be limited by the non-linearity of the state space for the rotation matrices $SO(3)$. It is this non-linearity that generates the highly non-linear terms in (14.40).

In this section, we are interested in analysing the performance of the designed control law in the presence of disturbances due to unknown air resistance forces. It is desirable to run a number of simulations to observe the effect of such disturbances on the behaviour of the helicopter. To this end, it would be possible to simulate the entire non-linear dynamics of the helicopter with added perturbation terms σ_t. There is, however, a simpler route available that is based on an error dynamic analysis of the closed-loop system. This analysis actually provides a better insight into the proposed control design. Let

$$\zeta = \left(e^T, m\alpha^T, \beta^T, \delta^T\right)^T \in \mathbb{R}^{15}$$

be the vector made up of all the various errors encountered during the backstepping procedure. The error coordinates ζ are a "linearizing" (for the closed-loop system) set of coordinates for the unperturbed system of (14.25)-(14.29). In fact, the error dynamics are only dimension 15 rather than 16 (the dimension of the augmented state for (14.25)-(14.29)). This is due to the fact that only the angular velocity of γ is measured as an error. The position of γ is not regulated, and indeed, continually increases as the rotor blades rotate. Even though the error dynamics do not completely represent the system dynamics, they contain all the information relevant to the analysis of the robustness of the closed-loop system to perturbations due to unmodelled air resistance terms.

It may be verified from the development in Section 14.3 that the closed-loop error dynamics in the absence of perturbation are linear

$$\dot{\zeta} = A\zeta$$

where

$$A := \begin{pmatrix} -I_3 & \frac{1}{m}I_3 & 0 & 0 & 0 & 0 & 0 \\ 0 & -\frac{1}{m}I_3 & I_3 & 0 & 0 & 0 & 0 \\ 0 & -\frac{1}{m}I_3 & -I_3 & 0 & I_3 & 0 & 0 \\ 0 & 0 & 0 & -1 & 0 & 1 & 0 \\ 0 & 0 & -I_3 & 0 & -rI_3 & 0 & 0 \\ 0 & 0 & 0 & -1 & 0 & -r & 0 \\ 0 & 0 & 0 & 0 & 0 & 0 & -r \end{pmatrix} \in \mathbb{R}^{15 \times 15}$$

Note that the bottom right hand 5×5 block (associated with the δ dynamics) is multiplied by a scale factor of r. Thus, one may impose a time scale separation property on the error dynamics, ensuring that the δ dynamics function evolve much more quickly than the other error dynamics. To see why this is necessary, one must study the influence of the perturbation σ_t on the error dynamics. The influence of the perturbation σ_t enters as an input disturbance into the linear error dynamics. Recalling (14.11) and (14.35) then it may be directly verified that the closed-loop error dynamics in the presence of the perturbation can be written as

$$\dot{\zeta} = A\zeta - B(\eta, u)\sigma_t$$

where $B(\eta, u) := \mathrm{diag}(I_{10}, K(\eta, u))e_{15} \in \mathbb{R}^{15}$. In particular, $B(\eta, u)$ is non-zero only in the last five entries corresponding to error δ. Thus, to reduce the effect of the air resistance perturbations, it is necessary to assign stable error dynamics for δ that dominate the perturbation effects. *An important point to make here is that it is not sufficient to simply dominate the perturbation effects observed in the regulation of the angular velocity $\dot{\gamma}$ of the main rotor.* Rather, it is necessary to stabilize the angular velocity of the full system dynamics despite the presence of the perturbation. This observation helps understand why many teams working on helicopters have had considerable trouble obtaining good control performance with a separate control loop regulating rotor velocity, no matter how tightly the rotor velocity is controlled.

To compute the matrix $B(\eta, u)$, it is necessary to compute the fifth column of $K(\eta, u)$ (14.35). This is given by

$$K(\eta, u)_{(\cdot, 5)} := -\frac{1}{c_\theta^2 I_3^a I_3^b} \bar{J}(\eta, u) \begin{pmatrix} -c_\theta c_\psi I_3^b \\ c_\theta^2 s_\psi I_3^b \\ -s_\theta c_\theta c_\psi I_3^b \\ 0 \\ c_\theta^2 (I_3^a + I_3^b) \end{pmatrix}$$

where the column vector is the fifth column of $S\mathrm{diag}\left(\mathbb{I}^{-1}, 1\right)$. Due to the permutation matrix S, this is just the fourth column of the inverse inertia matrix $\mathbb{I}^{-1}_{(\cdot, 4)}$ (cf. (14.16)) with its fourth entry swapped into the fifth entry of the above vector and a zero in the fourth entry.

From (14.30) and (14.33), it may be verified that

$$\bar{J}(\eta, u) = \begin{pmatrix} \bar{J}_{11} & u\left(c_\psi c_\theta c_\phi\right) & \bar{J}_{13} & \bar{J}_{14} & 0 \\ \bar{J}_{21} & u\left(c_\psi c_\theta s_\phi\right) & \bar{J}_{23} & \bar{J}_{24} & 0 \\ 0 & -uc_\psi s_\theta & -us_\psi c_\theta & \left(c_\psi c_\theta\right) & 0 \\ 1 & 0 & 0 & 0 & 0 \\ 0 & 0 & 0 & 0 & 1 \end{pmatrix}$$

where

$$
\begin{aligned}
\bar{J}_{11} &= u\left(s_\psi c_\phi - c_\psi s_\theta s_\phi\right)\\
\bar{J}_{13} &= u\left(c_\psi s_\phi - s_\psi s_\theta c_\phi\right)\\
\bar{J}_{14} &= \left(c_\psi s_\theta c_\phi + s_\psi s_\phi\right)\\
\bar{J}_{21} &= u\left(s_\psi s_\phi + c_\psi s_\theta c_\phi\right)\\
\bar{J}_{23} &= u\left(-c_\psi c_\phi - s_\psi s_\theta s_\phi\right)\\
\bar{J}_{24} &= \left(c_\psi s_\theta s_\phi - s_\psi c_\phi\right)
\end{aligned}
$$

Tedius but direct calculations yield

$$
B(\eta, u) = -\frac{1}{c_\theta^2 I_3^a I_3^b}
\begin{pmatrix}
0_{10}\\
\begin{pmatrix}
-uc_\theta c_\psi \left(s_\psi c_\phi - c_\psi s_\theta s_\phi\right) I_3^b - uc_\theta^3 s_\psi c_\psi c_\phi I_3^b\\
+us_\theta c_\theta c_\psi \left(c_\psi s_\phi - s_\psi s_\theta c_\phi\right) I_3^b
\end{pmatrix}\\
\begin{pmatrix}
-uc_\theta c_\psi \left(s_\psi s_\phi + c_\psi s_\theta c_\phi\right) I_3^b + uc_\theta^3 s_\psi c_\psi s_\phi I_3^b\\
-us_\theta c_\theta c_\psi \left(-c_\psi c_\phi - s_\psi s_\theta s_\phi\right) I_3^b
\end{pmatrix}\\
\left(-uc_\theta^2 s_\psi c_\psi s_\theta I_3^b + us_\theta c_\theta^2 c_\psi s_\psi I_3^b\right)\\
-\left(c_\theta c_\psi I_3^b\right)\\
c_\theta^2 (I_3^a + I_3^b)
\end{pmatrix}
$$

where $0_{10} \in \mathbb{R}^{10}$ is the zero vector with ten entries.

Apart from its evident complexity, there are two observations that are of interest to make for $B(\eta, u)$. Firstly, it is instructive to think of the situation in regards to a full scale helicopter where the inertia of the airframe strongly dominates that of the rotors, $I_A^3 >> I_3^b$. In this case, only the terms containing a component I_3^a contribute significantly to the matrix $B(\eta, u)$. It follows that only the last entry of $B(\eta, u)$ contributes significantly to $B(\eta, u)$. Consequently, the effect of the perturbation σ_t is restricted almost entirely to a variation in the γ dynamics. In this case, it is reasonable to compensate for the unknown perturbations due to rotor drag by using a simple decoupled SISO control loop to regulate the rotor speed. If, however, the ratio I_3^b/I_3^a is large, then all the non-zero entries of $B(\eta, u)$ will contribute non-trivial perturbations to the dynamics and a simple SISO loop regulating rotor speed will not make the overall control design more robust. In the case of model helicopters, the ratio $I_3^b/I_3^a \approx 0.2-0.5$ and in the example considered in this chapter, the ratio is roughly $I_3^b/I_3^a \approx 0.34$.

The second observation is that the control input u is an integral part of vector B. For most manoeuvres, the control u is of a magnitude equal to roughly the mass times gravitational acceleration, since it is directly related to the force that supports the helicopter in flight. In the example considered below, the control u is roughly of magnitude 180. The scaling of the perturbation by the control u introduces considerable problems, since even for small perturbations σ the actual effect on the error dynamics may dominate the desired dynamic response. To counteract the effect of the perturbations and to stop them propagating into the other error variables, it is necessary to scale the δ error dynamics to deal with the size perturbations encountered. In the proposed control design, this is achieved by introducing a scaling factor r related to the exponential rate of stability of the δ error dynamics. In the simulations that follow, we have chosen $r \approx 50$. Choosing r larger (of order of the magnitude of u) allows one to tolerate larger disturbances σ_t, however, imposes issues regarding achievability of control action (requiring more aggressive control action) and issues of robustness of the error dynamic modelling (given that the time scale separation is imposed and not natural to the system).

Parameter	Value
Mass	18.085 kg
I_1^a	1.667 kg m^2
I_2^a	2.341 kg m^2
I_3^a	1.197 kg m^2
Main Rotor diameter	1.798 m
Rotor blade mass	0.250 kg
I_3^b	0.404 kg m^2
r	50

Table 14.1: Parameters of helicopter used

Two simulations have been included to give an indication of the performance of the proposed control strategy in the presence of unmodelled rotor drag disturbances. The parameters of the helicopter considered are given in Table 14.1. For both simulations, the trajectory considered was a helical trajectory with ascending verticle height. The radius of the circle was 10 metres and the "yaw" trajectory was chosen in order that the helicopter would follow the trajectory as if in normal flight, i.e. flying along the helical trajectory with pilot facing forward. The initial condition for the helicopter was chosen to be the center of the helix at the correct height, ensuring that the initial error e and α of the Lyapunov function would be non-zero. For simplicity's sake, the higher

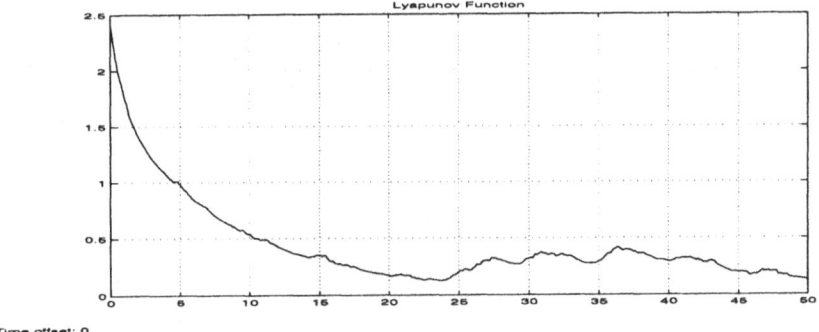

Figure 14.2: The decrease in the Lyapunov function for the case where $|\sigma_t| \approx 1$

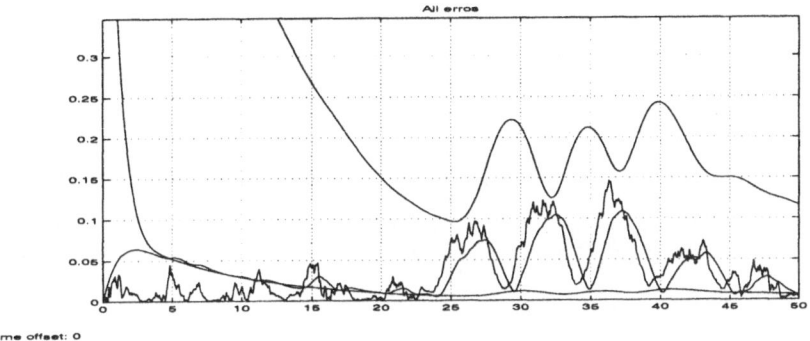

Figure 14.3: The norms of the error signals that comprise the Lyapunov function

order components of the Lyapunov function (β, δ) were set to zero to commence with. The disturbance σ_t was generated using filtered white noise signal multiplied by a gain. The same basic noise signal was used for both simulations and only the gain of the noise term was changed.

Two simulations have been included to show the effect of changing the gain on the perturbation σ_t, display the performance of the control algorithm, and provide a comparison with the results of [48]. The first simulation, displayed in Figures 14.2 to 14.5, was completed for unity gain. Thus

$$|\sigma_t| \approx 1$$

and Figure 14.5 is a true representation of the perturbation term. Figure

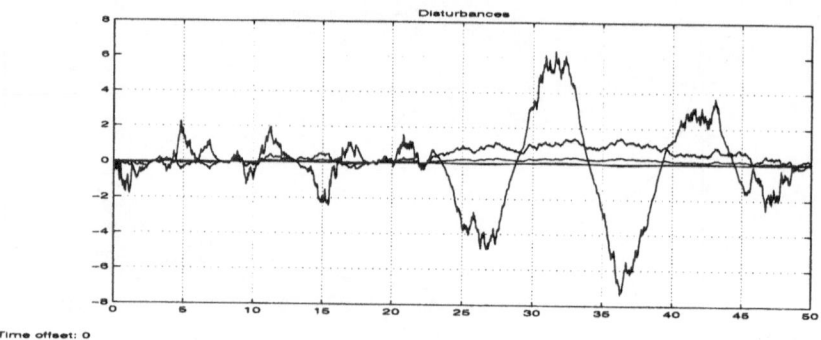

Figure 14.4: The values of the five non-zero entries of the perturbation term $B(\eta, u)\sigma_t$

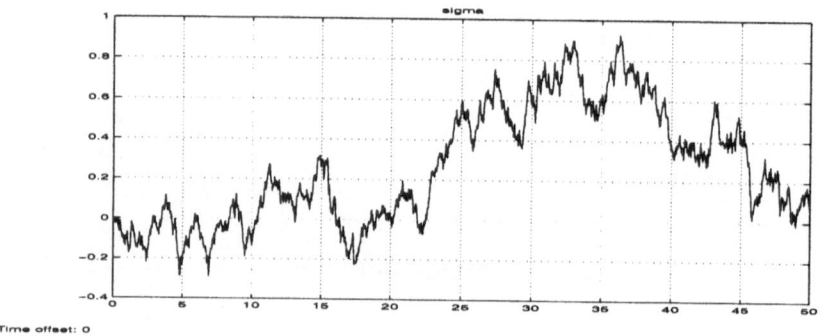

Figure 14.5: The value of σ_t during both simulations. The second simulation is undertaken for the actual perturbation signal $0.05 * \sigma_t$

14.2 shows the evolution of the Lyapunov function and indicates that there is certainly a measurable effect on the evolution of the system due to the perturbations. Nevertheless, the Lyapunov appears bounded by 0.5 and this ensures that the error $|\mathcal{E}| \leq 0.5$, or that the tracking performance should be better than 50cm. In fact, since the Lyapunov function measures all components of the error dynamics, one would expect that the actual position error would be much smaller than this. This is the case and is shown in Figure 14.3, which shows a scaled version of the evolution of each of the error signals ($|e|, |\alpha|, |\beta|, |\delta|$). The actual position error is the exponential error, which is non-zero to begin with and then appears to converge exponentially to zero. Thus, it appears

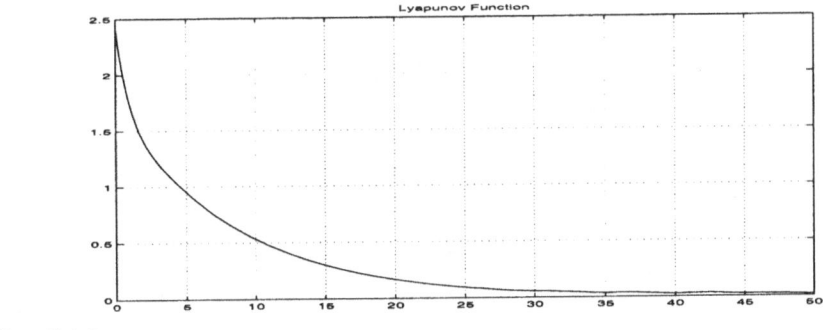

Figure 14.6: The decrease in the Lyapunov function for the second simulation where actual perturbation signal is $0.05\sigma_t$

that the actual position error is regulated to within an error $|e| \leq 0.05$ of 5cm. To give an indication of the actual perturbation terms, Figure 14.4 shows the evolution of the non-zero entries of $B(\eta, u)$.

The second simulation was undertaken for a gain of 0.05 multiplying the value of σ_t. Thus

$$|\sigma_t| \approx 0.05$$

This value has been chosen to compare to the results obtained in [48]. As can be seen in Figure 14.6, the resulting perturbations are effectively negligible and the Lyapunov function appears to be monotonically decreasing. Tracking is achieved to within an error of 1cm or less. Figure 14.7 provides a close-up of the error signals ($|e|, |\alpha|, |\beta|, |\delta|$) and indicates that the perturbation effects are still present in all the error terms, however, the magnitude of the perturbation is negligible.

Though the present simulations have not been verified on an actual helicopter, the fact that several different teams working in this area world wide have had difficulties with rotor velocity regulation indicates that this issue is important. The authors believe that the actual perturbations σ_t encountered in practical experiments will be non-negligible in real applications and thus the performance of the proposed control algorithm should be of interest in the design of robust controllers for model helicopters.

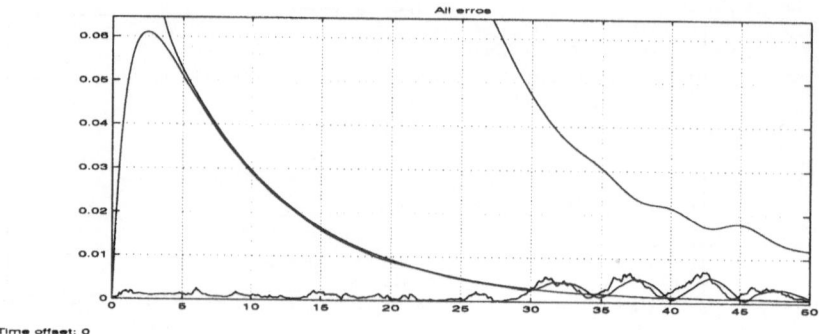

Figure 14.7: The values of the norms of the errors making up the Lyapunov function for the second simulation where the actual perturbation signal is $0.05\sigma_t$

14.5 Conclusions

In this chapter, we have shown that a Lagrangian model of the dynamics of a helicopter, simplified to avoid considering certain aerodynamic effects, permits a Lyapunov design of a unified path tracking control algorithm that combines position and orientation regulation with the regulation of the angular velocity of the rotors. The design method uses relative degree matching and dynamic extension to overcome a mismatch of inputs and desired tracking errors with the natural pure feedback structure of the system. The unified design procedure leads to a robust control design with respect to poorly modelled air resistance terms.

Chapter 15

Newtonian helicopter model

15.1 Introduction

In recent years, there has been growing interest within the control community on the subject of construction and control of autonomous model helicopters and experimental helicopter platforms [66, 90, 103, 119]*. It appears that the classical modelling and control approaches (cf. [87]) are not directly applicable due to the high actuation to inertia ratios and the highly non-linear nature of the rotation dynamics exploited in desired flight conditions for scale model autonomous helicopters. This has led the community to develop an idealised non-linear dynamic model for a scale model autonomous helicopter (cf. conference papers [28, 62, 90, 95, 101, 103, 119] and more recently the journal papers [93, 102]). Although the model that is becoming standard in the literature does not contain a sophisticated aerodynamic analysis and concerns only the basic dynamic states of the helicopter, it is hoped that by resolving the basic trajectory planning and control issues, it will be possible to extend these developments to provide robust practical controllers for scale model autonomous helicopters. The key techni-

*The authors of this chapter are Robert Mahony, Tarek Hamel, Alejandro Dzul and Rogelio Lozano. R. Mahony is with the Department of Electrical & Computer Systems Engineering, Monash University, Clayton, Victoria, 3800, Australia. T. Hamel is with the Cemif, Université d'Evry, 40 rue du Pelvoux, CE 1455 Courcouronnes, France. A. Dzul and R. Lozano are with the Laboratory Heudiasyc, UTC UMR CNRS 6599, Centre de Recherche de Royallieu, BP 20529, 60205 Compiègne Cedex, France.

cal difficulty encountered is the the presence of significant small body forces [12, 116] leading to weakly non-minimum phase zero dynamics for the full dynamic model [47]. This is a theoretical problem that was also encountered in the investigation of the control of a vertical take-off and landing jet (VTOL) [31, 35, 67, 73, 115] and [91, pg. 246.]. Unfortunately, the differential flatness technics applicable in the case of a VTOL do not apply in general to a helicopter [53, 64, 66, 116]. Recent work [102, 119] exploits the partial differential flatness properties that do exist for the helicopter model, however, the final stabilizing control design still relies on an approximation of the model. Most other authors have applied a robust control design to the model obtained by ignoring the small body forces and later analyzing the performance of the system to ensure that for desired trajectories, the unmodelled dynamics do not destroy the stability of the closed-loop system [28, 62, 71, 93]. Such results either take the form of monitoring the behaviour of the system in order to ascertain when the stability guarantees of the control design are broken (cf. Lemma 15.1) or provide some a priori guarantees for a restricted class of trajectories [28, 61, 93].

In this chapter, a detailed derivation of the standard non-linear model of the helicopter is presented. Rather than model the main rotor, tail rotor and airframe as a single mechanical system, their force and torque interactions are considered and the Newtonian dynamics of the airframe is presented. Using the standard model, a robust control design based on robust backstepping techniques [29] is proposed. A control Lyapunov function is derived, based on the block pure feedback form [49] of the approximate dynamic model obtained by ignoring the small body forces. The trajectory tracking control design is developed independently of a local coordinate parameterisation of the helicopter orientation, however, Euler angles are used to parameterise the final "yaw" parameter that does not explicitly contribute to the flight trajectory dynamics in the model considered. The Lyapunov function obtained for the closed-loop system is used to analyse the performance of the proposed control in tracking an arbitrary trajectory. A lemma is given (Lemma 15.1) that monitors the performance of the control relative to the decrease in the Lyapunov function on-line.

The chapter is arranged into six sections, including the present introduction. Section 15.2 presents the the general model of the dynamics of a scale model autonomous helicopter. Section 15.4 derives a Lyapunov control for an approximate system based on that presented in Section 15.2 via a robust backstepping approach. Section 15.5 contains the main results of the chapter and presents the analysis of the closed-loop performance of the proposed control. Section 15.6 presents a series of experiments that show the behaviour of the closed-loop system for both the approximate and complete dynamic models of the system.

15.2 Modeling a helicopter using Newton's laws

In this section, a dynamic system is proposed as a model for an autonomous model helicopter in flight close to hover conditions.

Consider Figure 15.1. Denote the body or airframe of the helicopter by the letter $\mathbf{A}$. Let $\mathcal{I} = \{E_x, E_y, E_z\}$ denote a right hand inertial frame stationary with respect to the earth and such that E_z denotes the vertical direction downwards into the earth. The vector $\xi = (x, y, z)$ denotes the position of the centre of mass of the helicopter relative to the frame $\mathcal{I}$. Let $\mathcal{A} = \{E_1^a, E_2^a, E_3^a\}$ be a (right hand) body fixed

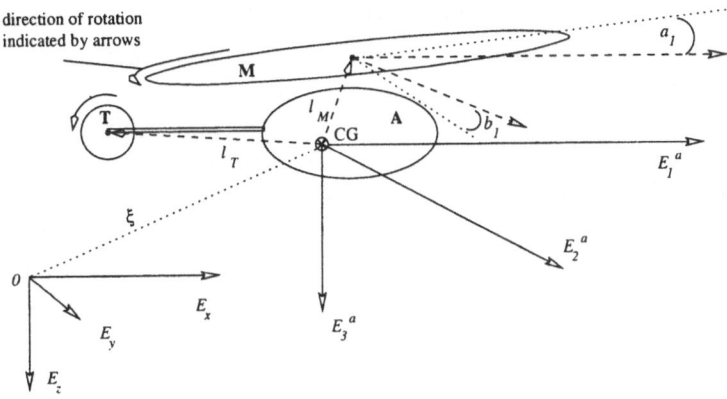

Figure 15.1: Geometry of model helicopter

frame for $\mathbf{A}$. We choose E_1^a to correspond to the normal direction of flight of the helicopter, E_2^a is orthogonal and in the horizontal plane, while E_3^a should (hopefully) correspond with E_z in normal stationary hover conditions (cf. Figure 15.1). The orientation of the helicopter is given by a rotation $R : \mathcal{A} \rightarrow \mathcal{I}$, where $R \in SO(3)$ is an orthogonal rotation matrix.

The aerodynamic and centrifugal forces acting on the rotor blades are more than 100 times stronger than the forces associated with the rigidity of the rotor blades. Consequently, for a non-zero cyclic pitch input (angle of attack of the rotor blades adjusted cyclically each rotation), the blades will quickly assume new trajectories that balance the aerodynamic and centrifugal forces. The response of the rotor blades to cyclic pitch inputs is effectively instantaneous (taking around 130° azimuth of a single rotation [87, pg. 462], corresponding to a time constant of approximately 0.02s for a reduced scale helicopter. The balance between aerodynamic and centrifugal forces occurs with the rotor blades lying in a disk, termed the main rotor disk, whose orientation is not fixed perpendicular to the hub axis of the rotor blades. Nevertheless, controlling the cyclic pitch of the rotor blades results in torque control over pitch and roll of the airframe, since tilting the main rotor disk effects the orientation of the main lift force and generates rotational torque around the centre of mass of the helicopter. The rotational torque obtained is directly proportional to the horizontal component of the main rotor thrust and the offset between the main rotor hub and the centre of mass. In addition to causing a rotation, the horizontal component of the main rotor thrust results in small sideways forces applied to the helicopter airframe. These forces are termed as small body forces.

Clearly, the tilt of the main rotor disk is a key parameter in the representation of the dynamics of a helicopter. The orientation of the main rotor disk is represented classically [87] by two additional angles a_1 and b_1, representing the longitudinal and lateral (respectively) tilt of the rotor disk with respect to the airframe **A**. The angles are often termed the flapping angles, since the tilt of the main rotor disk is associated with a vertical flapping movement of the rotor blades while they rotate. Thus, the longitudinal flapping angle, a_1, measures the deflection of the rotor disk associated with a rotation around E_2^a, while the latitudinal flapping angle, b_1, measures a deflection of the rotor disk associated with a rotation around E_1^a (cf. Figure 15.1).

Assumption 15.1 *The main rotor blades are assumed to hinge directly from the hub, that is, there is no hinge offset associated with rotor flapping. The coning angle is not considered. It is assumed that the cyclic longitudinal and lateral tilts of the main rotor disk are measurable and controllable via control of the cyclic pitch and these angles are taken directly as control inputs. The only air resistance considered is simple drag forces opposing the rotation of the two rotors.* ∎

Consider the contributions to the helicopter's motion from lift associated with the rotor disks. Denote the thrust forces for the main rotor and tail rotor by T_M and T_T respectively. These forces will act to provide both a translation force and a rotational torque due to their offset from the centre of gravity. The tail rotor has a fixed orientation and the thrust generated may be represented by $T_T = T_T^2 E_2^a = (0, T_T^2, 0)^T$ in the body fixed frame $\mathcal{A}$. The orientation of the main rotor thrust T_M is normally expressed in terms of the lateral and longitudinal cyclic tilt angles a_1 and b_1 (cf. Figure 15.2). The main rotor thrust may be

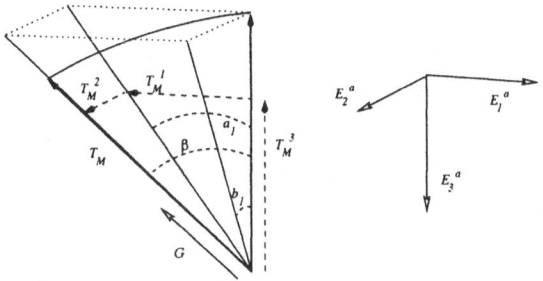

Figure 15.2: Orientation of the thrust vector T_M and definition of the unit vector $G := G(a_1, b_1)$. Here, β (the flapping angle) denotes the maximal tilt of the rotor disk (and thus the maximal angle of vertical flapping of each rotor during its cycle)

expressed as a vector in the body fixed frame $\mathcal{A}$ as follows

$$T_M = G(a_1, b_1)|T_M|$$

where

$$G(a_1, b_1) = \frac{1}{d(a_1, b_1)} \begin{pmatrix} -\sin(a_1)\cos(b_1) \\ \sin(b_1)\cos(a_1) \\ -\cos(a_1)\cos(b_1) \end{pmatrix} \qquad (15.1)$$

$(d(a_1, b_1) := \sqrt{1 - \sin^2(a_1)\sin^2(b_1)})$ is the unit vector in the direction of the main rotor lift (cf. Figure 15.2).

The total translational force applied to the airframe $\mathbf{A}$ expressed in the inertial frame $\mathcal{I}$ is

$$f = RG(a_1, b_1)|T_M| + T_T^2 Re_2 + mge_3 \qquad (15.2)$$

where the gravitational force $mge_3 \in \mathcal{I}$ is added and the force contribution T_M and T_T discussed above are translated into the inertial frame via the rotation $R : \mathcal{A} \to \mathcal{I}$.

The two thrust vectors T_M and T_T also generate torques, τ_M and τ_T, due to the respective offsets l_M and l_T (cf. Figure 15.1) between the centre of mass and the rotor hubs

$$\tau_M = l_M \times T_M = |T_M| l_M \times G(a_1, b_1)$$
$$\tau_T = l_T \times T_T = T_T^2 l_T \times E_2^a$$

In addition to the forces and torques produced by the thrust forces of the rotors, there are pure torques acting through the rotor hubs associated with the reactive torque generated by the aerodynamic drag on the rotors. These anti-torques are denoted Q_M and Q_T for the main and tail rotors respectively and act along the axis of the rotor hubs. It follows that the anti-torque on the airframe is

$$Q_M = |Q_M| E_3^a, \quad Q_T = -|Q_T| E_2^a$$

Denote the total torque applied to the airframe **A** expressed in the body fixed frame by τ. Thus

$$\begin{aligned} \tau = &|T_M| \left[l_M \times G(a_1, b_1) \right] + T_T^2 \left[l_T \times E_2^a \right] \\ &+ |Q_M| E_3^a - |Q_T| E_2^a \end{aligned} \tag{15.3}$$

Newton's classical equations of motion for a rigid object evolving in $SE(3)$ directly yield a dynamic model for a reduced scale helicopter, where the force inputs are given by the expressions derived above

$$\dot{\xi} = v, \quad m\dot{v} = f \tag{15.4}$$
$$\dot{R} = R\mathrm{sk}(\Omega), \quad \mathbf{I}\dot{\Omega} = -\Omega \times \mathbf{I}\Omega + \tau \tag{15.5}$$

where m is the total mass of the helicopter, $\mathbf{I}$ is the inertia of the airframe around the centre of mass and $\mathrm{sk} : \mathbb{R}^3 \to \mathbb{R}^{3\times3}$ takes a vector v to the associated skew-symmetric matrix such that $\mathrm{sk}(v)w = v \times w$.

15.3   New dynamic model for control design

In this section, the dynamic model given by (15.4)-(15.5) is considered and a new (equivalent) model is derived that is more convenient to work with for the purposes of control design.

Consider solely those forces and torques engendered by the rotors. In the translation dynamics (15.4), the rotor forces are

$$F_r = |T_M| RG(a_1, b_1) + T_T^2 Re_2 \tag{15.6}$$

In the rotation dynamics equation (15.5), the torque contributions from
the rotors are

$$
\tau_r = |T_M| \begin{pmatrix} 0 & -l_M^3 & l_M^2 \\ l_M^3 & 0 & -l_M^1 \\ -l_M^2 & l_M^1 & 0 \end{pmatrix} G(a_1, b_1) + \begin{pmatrix} -l_T^3 \\ 0 \\ l_T^1 \end{pmatrix} T_T^2 \quad (15.7)
$$

where $l_M = (l_M^1, l_M^2, l_M^3)$ and $l_T = (l_T^1, l_T^2, l_T^3)$ are the component repre-
sentations of l_M and l_T in $\mathcal{A}$. In (15.7), taking the first two columns of
the first term along with the second term into a single matrix yields

$$
\tau_r = K \begin{pmatrix} |T_M| G^1(a_1, b_1) \\ |T_M| G^2(a_1, b_1) \\ T_T^2 \end{pmatrix} + k_0 |T_M| G^3
$$

where

$$
K = \begin{pmatrix} 0 & -l_M^3 & -l_T^3 \\ l_M^3 & 0 & 0 \\ -l_M^2 & l_M^1 & l_T^1 \end{pmatrix}, \quad k_0 = \begin{pmatrix} l_M^2 \\ -l_M^1 \\ 0 \end{pmatrix} \quad (15.8)
$$

Think of the term involving K as the one that contributes the control
over the rotation dynamics, while the term involving k_0 is a coupling
between the translation force control and the rotation dynamics. It is
natural to introduce a set of nominal control inputs $w = (w^1, w^2, w^3)^T$
providing rotational control around the body fixed frame coordinate
axis via

$$
w = K \begin{pmatrix} |T_M| G^1(a_1, b_1) \\ |T_M| G^2(a_1, b_1) \\ T_T^2 \end{pmatrix} \approx \begin{pmatrix} -a_1 |T_M| \\ b_1 |T_M| \\ T_T^2 \end{pmatrix} \quad (15.9)
$$

where the approximation $G^1(a_1, b_1) \approx -a_1$ and $G^2(a_1, b_1) \approx b_1$ results
from applying the small angle assumption to the trigonometric functions
in (15.1). It is important to note that $|l_M^3| >> |l_M^1|$, $|l_M^2|$ and $|l_T^1| >>$
$|l_T^2|, |l_T^3|$. Thus, the matrix K is clearly full rank. Moreover, if $|T_M| >>$
0 is large (as expected in normal flight conditions), then small changes
in the cyclic tilt angles a_1 and b_1 along with the tail thrust T_T^2 will allow
arbitrary (though bounded) control action w.

Consider the translation force (15.6). Recall that the rotation matrix
R may be written $R = [E_1^a \ E_2^a \ E_3^a]$, where the vectors E_i^a are expressed
in inertial frame $\mathcal{I}$ and represent the orientation of the body fixed frame

$\mathcal{A}$. One may write

$$F_r = E_3^a |T_M| G^3(a_1, b_1) + [E_1^a \; E_2^a \; E_2^a] \begin{pmatrix} |T_M| G^1(a_1, b_1) \\ |T_M| G^2(a_1, b_1) \\ T_T^2 \end{pmatrix}$$

Define a nominal control, $u > 0$, associated with the lift force due to the main rotor

$$u = -|T_M| G^3(a_1, b_1) = T_M^3 \tag{15.10}$$

as the component of the principal thrust in the direction E_3^a. Let $L = [e_1, e_2, e_2]$ (where e_i is the unit vector with a one in the ith place) and note that $[E_1^a \; E_2^a \; E_2^a] = RL$. Moreover, $(E_3^a)^T RL = e_3^T L = 0$. Substituting from (15.9) and (15.10) and recalling that K is invertible yields

$$F_r = -u Re_3 + RLK^{-1}w \tag{15.11}$$

Thus, the translation force can be written as two terms: the first is the principal control input for the translation dynamics. It is always directed in the negative E_3^a axis corresponding to the orientation of the airframe and position control must be achieved by reorienting the airframe. The second term is orthogonal to E_3^a and is expected to be of a much smaller magnitude. It corresponds to the small body forces exerted on the airframe when torque control is applied. Recalling the development given above for the rotational torques and substituting from (15.9) and (15.10), one obtains

$$\tau_r = w + k_0 u$$

This equation clearly shows the full torque control available via the control w and the separate term due to the coupling between translational and rotation inputs expressed as $k_0 u$.

Remark 15.1 *The above representation for the rotational torque applied to the airframe is not strictly true as a model of the cyclic pitch inputs to applied torque due to parasitic effects resulting from the mechanical properties of the rotor blades. In particular, the rigidity of the rotor blades results in a small contribution to the torque applied to the airframe (when the rotor disk deforms). More importantly, the rotor blade rigidity alters the dynamics of the flapping response of the rotor blades, leading to a small phase offset between the applied cyclic pitch* *direction and the deformation of the main rotor disk. Furthermore, there are effects such as the Coriolis force associated with the effective flapping hinge that contributes to total torque applied to the airframe. Finally, all scale model helicopters are equipped with stabilizer rotors to slow down the dynamic response to environmental disturbances. A full modelling of these effects is beyond the present development and the approach taken is equivalent to working with the flapping angles directly. This is perhaps not such a bad idea in practice, if a suitable method of measuring the flapping angles is developed, since the highly non-linear effects associated with the rotor flapping dynamics is one of the major blocks to effect control of scale model autonomous helicopters.* ∎

From the above development, one obtains the following equations of motion for a model helicopter.

$$\dot{\xi} = v \tag{15.12}$$

$$m\dot{v} = -uRe_3 + mge_3 + RLK^{-1}w \tag{15.13}$$

$$\dot{R} = R\mathrm{sk}(\Omega) \tag{15.14}$$

$$\mathbf{I}\dot{\Omega} = -\Omega \times \mathbf{I}\Omega + |Q_M|e_3 - |Q_T|e_2 + w + k_0u \tag{15.15}$$

15.4 Lyapunov-based tracking control design

In this section, the structure of Equations (15.12)-(15.15) is considered. A natural choice to derive a control law for such a system is to exploit the procedure of backstepping. Results from the analysis of a control law of this nature are presented.

Figure 15.3 gives a block diagram representation of (15.12)-(15.15) as an input-output system from inputs (u, w) to the position ξ. Note the cascade of rotation dynamics into the input of the translation dynamics via the block denoted coupling which is the term $-uRe_3$ in (15.13).

The coupling terms $RLK^{-1}w$ and k_0u result in feed-forward and feedback (type) connections, which destroy the pure cascade structure of the system. The air resistance terms are represented as input disturbances to the rotational dynamics.

Consider the model equations (15.12)-(15.15). We wish to fully determine trajectories determining the evolution of the helicopter. The position trajectory is a simple matter to assign. To achieve a given trajectory, it will be necessary to manipulate the direction in which the principal translation force u acts and this will in turn determine trajectories for the pitch and roll of the helicopter. The yaw, however, is

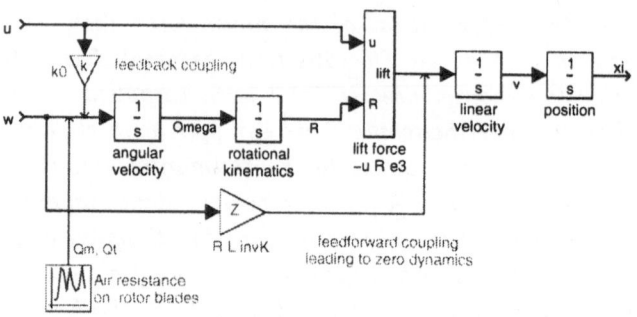

Figure 15.3: Cascade structure of Equations 15.12-15.15

also a free variable. To assign a trajectory to the helicopter yaw, it is necessary to measure the yaw in some manner. To do this, we introduce the Euler angles

$$\eta = (\phi, \theta, \psi) \tag{15.16}$$

which are the classical "yaw", "pitch" and "roll" Euler angles commonly used in aerodynamic applications [33, pg. 608]. Firstly, a rotation of angle ϕ around the axes E_z is applied, corresponding to "yaw". Secondly, a rotation of angle θ around the rotated version of the E_y axis is applied, corresponding to "pitch" of the airframe. Lastly, a rotation of angle ψ around the axes E_1^a is applied. This corresponds to "roll" of $\mathbf{A}$ around the natural axis E_1^a. It should be noted that the Euler angles $\eta = (\phi, \theta, \psi)$ are not a global coordinate patch on $SO(3)$. Indeed, when $\theta \geq \frac{\pi}{2}$, then the correspondence between the Euler coordinates and the rotation matrices in $SO(3)$ is no longer one-to-one. Using Euler angles to represent the system dynamics will not be a problem for manoeuvres close to hover. The rotation matrix $R := R(\phi, \theta, \psi) \in SO(3)$ representing the orientation of the airframe $\mathbf{A}$ relative to a fixed inertial frame may be written[1] in terms of the Euler angles $\eta = (\phi, \theta, \psi)$

$$R = \begin{pmatrix} c_\theta c_\phi & s_\psi s_\theta c_\phi - c_\psi s_\phi & c_\psi s_\theta c_\phi + s_\psi s_\phi \\ c_\theta s_\phi & s_\psi s_\theta s_\phi + c_\psi c_\phi & c_\psi s_\theta s_\phi - s_\psi c_\phi \\ -s_\theta & s_\psi c_\theta & c_\psi c_\theta \end{pmatrix} \tag{15.17}$$

---

[1]The following shorthand notation for trigonometric function is used:

$$c_\beta := \cos(\beta), \quad s_\beta := \sin(\beta)$$

Let

$$\hat{\xi} : \mathbb{R} \to \mathbb{R}^3$$

$$\hat{\phi} : \mathbb{R} \to \mathbb{R}$$

be smooth trajectories $\hat{\xi}(t) := (\hat{x}(t), \hat{y}(t), \hat{z}(t))$ and $\hat{\phi}(t)$. The control problem considered is:

Find a feedback control action (u, w^1, w^2, w^3) depending only on the measurable states $(\xi, \dot{\xi}, \dot{\eta}, \eta)$ and arbitrarily many derivatives of the smooth trajectory $(\hat{\xi}(t), \hat{\phi}(t))$ such that the tracking error

$$\mathcal{E} := (\xi(t) - \hat{\xi}(t), \phi(t) - \hat{\phi}(t)) \in \mathbb{R}^4 \tag{15.18}$$

is asymptotically stable for all initial conditions.

Define a partial error

$$\delta_1 := \xi(t) - \hat{\xi}(t) \tag{15.19}$$

comprising that part of the tracking error associated with the position coordinates. Define a first storage function

$$S_1 = \frac{1}{2}\delta_1^T \delta_1 = \frac{1}{2}|\delta_1|^2 \tag{15.20}$$

Taking the time derivative of S_1 and substituting for (15.12) yields

$$\frac{d}{dt}S_1 = \delta_1^T(\dot{\xi} - \dot{\hat{\xi}})$$
$$= \delta_1^T(v - \hat{v})$$

where $\hat{v} := \dot{\hat{\xi}}$. Let v_d denote a desired value for the velocity v. This theoretical signal is considered as a control signal [49] entering in place of the true velocity v and is chosen to ensure the storage function S_1 decreases,

$$v_d := \hat{v} - \frac{1}{m}\delta_1$$

With this choice, one has

$$\dot{S}_1 = \frac{1}{m}\delta_1^T(mv_d - m\hat{v}) + \frac{1}{m}\delta_1^T(mv - mv_d)$$
$$- \frac{1}{m}|\delta_1|^2 + \frac{1}{m}\delta_1^T(mv - mv_d)$$

The process of backstepping continues by considering a new error

$$\delta_2 := mv - mv_d$$

the difference between the desired and true velocities. Thus, the derivative $\dot{S}_1$ may be written as

$$\dot{S}_1 = -\frac{1}{m}|\delta_1|^2 + \frac{1}{m}\delta_1^T\delta_2$$

The second storage function considered is

$$S_2 = \frac{1}{2}|\delta_2|^2 = \frac{1}{2}|mv - mv_d|^2$$

Deriving S_2 and recalling (15.4) yields

$$\dot{S}_2 = \delta_2^T(m\dot{v} - m\dot{v}_d),$$
$$= \delta_2^T(-E_3^a u + mge_3 - m\dot{v}_d),$$
$$= \delta_2^T(-uR(\eta)e_3 + mge_3 - m\dot{v}_d)$$

where it is recalled that $E_3^a = R(\eta)e_3$. The backstepping process is continued with respect to the new variables (η, u). Of course the control u is available and could be directly assigned at this point. In preference to such an approach, it is possible to consider a dynamic extension of control u by a double integrator

$$\ddot{u} = \tilde{u} \qquad\qquad (15.21)$$

Thus, both the actual control u and its first derivative $\dot{u}$ become internal variables of a dynamic controller. The advantage of this process is that the relative degree of the new control, $\tilde{u}$, with respect to the position variables, ξ, is four rather than two. This agrees with the relative degree of the torque controls, w, with respect to the position variables, ξ. Matching relative degrees of the inputs relative to the considered outputs allows a combined assignment of the full dynamics of a higher order error term depending on the position coordinates ξ and additional coupling terms that are generated by the backstepping procedure. A block control assignment of this nature is better than a cascaded design, where u is assigned directly, since it more naturally allows trade-off between the various control objectives. In contrast, if u is assigned at this stage, it results in aggressive control for the translation (main rotor lift control) compared to the control of the rotation. The approach taken also results in algebraically less complicated equations.

Let (η_d, u_d) denote the desired values of η and the control u and set

$$u_d R(\eta_d) e_3 := mg e_3 - m\dot{v}_d + \delta_2 + \frac{1}{m}\delta_1$$

This assignment does not uniquely define the values of η_d and u_d. This is not a problem as the assignment fixed here is an element by element vectorial assignment of $u_d R(\eta_d) e_3$. Intuitively, one is specifying a desired direction and magnitude for the thrust associated with the main rotor. It is clear that any such vector may be assigned, since $R(\eta_d) e_3$ is an arbitrary rotation of a unit vector and u_d provides control of the magnitude. Indeed, the separate desired inputs η_d and u_d need never be considered independently and it is advantageous to introduce a notation to represent the desired vector input

$$X_d := u_d R(\eta_d) e_3 = mg e_3 - m\dot{v}_d + \delta_2 + \frac{1}{m}\delta_1 \qquad (15.22)$$

In later stages of the backstepping process, it is necessary to take derivatives of the desired vector direction X_d. These derivatives are computed analytically by differentiating the explicit expression on the right hand side. One need never actually compute the value or the derivatives of u_d and η_d. With the above choice, one has

$$\dot{S}_2 = -|\delta_2|^2 - \frac{1}{m}\delta_2^T \delta_1 + \delta_2^T (X_d - u R(\eta) e_3)$$

The process of backstepping continues by considering a third error

$$\delta_3 = u_d R(\eta_d) e_3 - u R(\eta) e_3 = X_d - u R(\eta) e_3 \qquad (15.23)$$

the vectorial difference between the desired and true values of translation thrust $u R(\eta) e_3$ and a further component, which penalizes the yaw

$$\epsilon_3 = \phi - \hat{\phi}$$

The yaw component of the error term is introduced at this stage of the backstepping procedure (rather than along with the initial error term δ_1) in order that the relative degree of δ_3 and ϵ_3 with respect to the controls $\tilde{u}$ and w match. Indeed, the relative degree of each control with respect to either error is two. With the choice of errors, δ_3 and ϵ_3, the derivative of S_2 may be written as

$$\dot{S}_2 = -|\delta_2|^2 - \frac{1}{m}\delta_2^T \delta_1 + \delta_2^T \delta_3$$

Consider the storage function

$$S_3 = \frac{1}{2}|\delta_3|^2 + \frac{1}{2}|\epsilon_3|^2$$

Deriving S_3 and recalling (15.14) yields

$$\dot{S}_3 = \delta_3^T(\dot{X}_d - (\dot{u}R(\eta)e_3 + R(\eta)\text{sk}(\Omega)e_3)) + \epsilon_3(\dot{\phi} - \dot{\phi}) \qquad (15.24)$$

Consider the term associated with δ_3 firstly. Let $(\Omega_d, \dot{u}_d)$ denote the desired values of Ω and the control derivative $\dot{u}$. Analogously to the case for $uR(\eta)e_3$, the full vectorial term

$$\dot{u}_dR(\eta)e_3 + R(\eta)\text{sk}(\Omega_d)e_3 := \dot{X}_d + \delta_3 + \delta_2 \qquad (15.25)$$

is assigned. Note that in this case, the value of $R(\eta)$ depends on the true value of η and not on a desired value. Thus, it is important that the above equation can always be solved, for suitable $\dot{u}_d$ and Ω_d, given an arbitrary vector on the right hand side. Note that

$$\text{sk}(\Omega_d)e_3 = \Omega_d \times e_3 = -\text{sk}(e_3)\Omega_d$$

Thus, since $\text{sk}(e_3)$ is rank two with entries only in the first and second columns, (15.25) may be written as

$$\begin{pmatrix} 0 & 1 & 0 \\ -1 & 0 & 0 \\ 0 & 0 & 1 \end{pmatrix} \begin{pmatrix} \Omega_d^1 \\ \Omega_d^2 \\ \dot{u}_d \end{pmatrix} := R(\eta)^T\left(\dot{X}_d + \delta_3 + \delta_2\right) \qquad (15.26)$$

It is clear then that the full desired dynamics for the right hand side may be assigned using only the desired signals Ω_d^1, Ω_d^2 and $\dot{u}_d$. This leaves Ω_d^3 free to control the yaw ϕ. Analogously to the case for $uR(\eta)e_3$, once the validity of the vectorial assignment has been determined, one need never again work directly with the object $\dot{u}_dR(\eta)e_3 + R(\eta)\text{sk}(\Omega_d)e_3$. Instead, we introduce a vector notation Y_d to represent this value

$$Y_d := \dot{u}_dR(\eta)e_3 + R(\eta)\text{sk}(\Omega_d)e_3 = \dot{X}_d + \delta_3 + \delta_2 \qquad (15.27)$$

Now consider the term associated with ϵ_3 in (15.24). The desired input into this term will be the yaw velocity of the helicopter $\dot{\phi}$. Let $\dot{\phi}_d$ denote the desired yaw velocity and choose

$$\dot{\phi}_d := \dot{\phi} - \epsilon_3$$

Again it is important to verify that this equation can be satisfied in practice. This is not immediately clear, since the actual signals that are involved in the next stage of the backstepping procedure are the signals Ω^1, Ω^2, Ω^3 and $\dot{u}$. The previous analysis showed that the desired angular velocity Ω_d^3 is the only free variable available to control $\dot{\phi}_d$. It is necessary to show that this is possible.

To proceed, it is necessary to recap the kinematic relationship between the Euler angles and the angular velocity of a rigid body. Such calculations can be found in most texts on classical mechanics (cf. Goldstein [33]). The generalized velocities $\dot{\eta} = (\dot{\phi}, \dot{\theta}, \dot{\psi})$ are related to the angular velocity Ω via the standard kinematic relationship [33, pg. 609]

$$\dot{\eta} = \frac{1}{\cos(\theta)} \begin{pmatrix} 0 & s_\psi & c_\psi \\ 0 & c_\theta c_\psi & -c_\theta s_\psi \\ c_\theta & s_\theta s_\psi & s_\theta c_\psi \end{pmatrix} \Omega = W_\eta^{-1} \Omega \qquad (15.28)$$

where

$$W_\eta := \begin{pmatrix} -s_\theta & 0 & 1 \\ c_\theta s_\psi & c_\psi & 0 \\ c_\theta c_\psi & -s_\psi & 0 \end{pmatrix} \qquad (15.29)$$

Replacing $\dot{\eta}$ by $\dot{\eta}_d$ and Ω by Ω_d one obtains

$$\dot{\eta}_d = \frac{1}{\cos(\theta)} \begin{pmatrix} 0 & s_\psi & c_\psi \\ 0 & c_\theta c_\psi & -c_\theta s_\psi \\ c_\theta & s_\theta s_\psi & s_\theta c_\psi \end{pmatrix} \Omega_d$$

Note that the kinematic matrix W_η still depends on the true variables η, not the desired variables. Multiplying by e_1^T, one obtains

$$\dot{\phi}_d = \frac{s_\psi}{c_\theta} \Omega_d^2 + \frac{c_\psi}{c_\theta} \Omega_d^3 \qquad (15.30)$$

Thus, in order that $\dot{\phi}_d$ is dependent on the free variable Ω_d^3, it is necessary that $0 < \frac{c_\psi}{c_\theta} < \infty$, or equivalently that both

$$\theta, \psi \in \left(\frac{-\pi}{2}, \frac{\pi}{2} \right) \qquad (15.31)$$

evolve within the open interval $(-\pi/2, \pi/2)$. This corresponds to the helicopter never turning upside down, a situation that would require a more sophisticated understanding of dynamics of a helicopter than that presented in the present chapter. Thus, for the purposes of the present chapter, we simply assume that (15.31) is valid at all times. With the choices made above, one may rewrite (15.24) as

$$
\begin{aligned}
\dot{S}_3 \;=\; & -|\delta_3|^2 - \delta_3^T \delta_2 - \epsilon_3^2 + \epsilon_3(\dot{\phi} - \dot{\phi}_d) \\
& + \delta_3^T (Y_d - (\dot{u}R(\eta)e_3 + uR(\eta)\mathrm{sk}(\Omega)e_3))
\end{aligned}
\tag{15.32}
$$

For the next and last stage of the backstepping algorithm, one considers the new error terms

$$
\delta_4 = Y_d - (\dot{u}R(\eta)e_3 + uR(\eta)\mathrm{sk}(\Omega)e_3)
\tag{15.33}
$$
$$
\epsilon_4 = \dot{\phi} - \dot{\phi}_d
$$

With this choice, the derivative of S_3 may be written as

$$
\dot{S}_3 = -|\delta_3|^2 - \delta_3^T \delta_2 - \epsilon_3^2 + \delta_3^T \delta_4 + \epsilon_3 \epsilon_4
$$

The storage function associated with this stage of the backstepping is

$$
S_4 = \frac{1}{2}|\delta_4|^2 + \frac{1}{2}|\epsilon_4|^2
$$

Thus, taking the derivative of S_4 yields

$$
\begin{aligned}
\dot{S}_4 \;=\; & \delta_4^T (\dot{Y}_d - (\ddot{u}R(\eta)e_3 + 2\dot{u}R(\eta)\mathrm{sk}(\Omega)e_3 \\
& + uR(\eta)(\dot{\Omega} \times e_3))) + \epsilon_4(\ddot{\phi} - \ddot{\phi})
\end{aligned}
\tag{15.34}
$$

At this stage, the actual control inputs enter into the equations through $\ddot{u} = \tilde{u}$, $\dot{\Omega}$ via (15.15) and $\ddot{\phi}$ as seen below.

To simplify the following analysis, consider the following control input transformation of (15.15). Define

$$
\begin{aligned}
\tilde{w} \;:=\; & -\mathbf{I}^{-1}\Omega \times \mathbf{I}\Omega + |Q_M|\mathbf{I}^{-1}e_3 - |Q_T|\mathbf{I}^{-1}e_2 \\
& + \mathbf{I}^{-1}w + \mathbf{I}^{-1}k_0 u
\end{aligned}
\tag{15.35}
$$

Since $\mathbf{I}$ is full rank, then this is certainly a bijective control input transformation between w and $\tilde{w}$. With this choice, (15.15) becomes

$$
\dot{\Omega} = \tilde{w}
\tag{15.36}
$$

Taking a second derivative of (15.28) yields

$$
\begin{aligned}
\ddot{\eta} &= -W_\eta^{-1}\dot{W}_\eta W_\eta^{-1}\Omega + W_\eta^{-1}\dot{\Omega} \\
&= -W_\eta^{-1}\dot{W}_\eta W_\eta^{-1} + W_\eta^{-1}\tilde{w}
\end{aligned}
$$

Thus, recalling (15.30) and multiplying by e_1^T, one obtains

$$\ddot{\phi} = -e_1^T W_\eta^{-1} \dot{W}_\eta W_\eta^{-1} \Omega + W_\eta^{-1} \dot{\Omega}$$
$$= -e_1^T W_\eta^{-1} \dot{W}_\eta W_\eta^{-1} \Omega + \frac{s_\psi}{c_\theta} \tilde{w}^2 + \frac{c_\psi}{c_\theta} \tilde{w}^3 \qquad (15.37)$$

Now, (15.34) may be rewritten as

$$\dot{S}_4 = \delta_4^T (\dot{Y}_d - 2\dot{u} R(\eta) \text{sk}(\Omega) e_3 - (\tilde{u} R(\eta) e_3 - u R(\eta) \text{sk}(e_3) \tilde{w}))$$
$$+ \epsilon_4 (\ddot{\phi} - \ddot{\hat{\phi}})$$

To achieve the desired control, choose

$$\tilde{u} R(\eta) e_3 - u R(\eta) \text{sk}(e_3) \tilde{w} = \dot{Y}_d - 2\dot{u} R(\eta) \text{sk}(\Omega) e_3 + \delta_3 + \delta_4 \qquad (15.38)$$
$$\ddot{\phi} = \ddot{\hat{\phi}} - \epsilon_4 - \epsilon_3 \qquad (15.39)$$

With these choices, the derivative of S_4 may be rewritten as

$$\dot{S}_4 = -|\delta_4|^2 - \delta_4^T \delta_3 - |\epsilon_4|^2 - \epsilon_4 \epsilon_3$$

It remains to show that (15.38) and (15.39) can be satisfied simultaneously. Rewriting (15.38) analogously to (15.26)

$$\begin{pmatrix} 0 & u & 0 \\ -u & 0 & 0 \\ 0 & 0 & 1 \end{pmatrix} \begin{pmatrix} \tilde{w}^1 \\ \tilde{w}^2 \\ \tilde{u} \end{pmatrix} = \qquad (15.40)$$
$$R(\eta)^T \left(\dot{Y}_d - 2\dot{u} R(\eta) \text{sk}(\Omega) e_3 + \delta_3 + \delta_4 \right)$$

Thus, as long as $u \neq 0$, the control signals $\tilde{w}^1$, $\tilde{w}^2$ and $\tilde{u}$ are uniquely determined by this equation. To obtain $\tilde{w}^3$, one solves for (15.37), yielding

$$\frac{c_\psi}{c_\theta} \tilde{w}^3 = \ddot{\hat{\phi}} - \epsilon_4 - \epsilon_3 + e_1^T W_\eta^{-1} \dot{W}_\eta W_\eta^{-1} \Omega - \frac{s_\psi}{c_\theta} \tilde{w}^2 \qquad (15.41)$$

All terms on the right hand side of this equation are known at this stage and $\tilde{w}^3$ is well defined as long as (15.31) is valid.

The above process has sufficed to fully define the control inputs $\tilde{w}^1$, $\tilde{w}^2$, $\tilde{w}^3$ and $\tilde{u}$. Using (15.21) and (15.35), one recovers the original control inputs u and w, which of course are themselves functions of more primitive variables of the systems including flapping angles and thrust components. The backstepping procedure achieves the monotonic decrease of the following Lyapunov function

$$\mathcal{L} = S_1 + S_2 + S_3 + S_4$$

This is easily verified by computing

$$\dot{\mathcal{L}} = \dot{S}_1 + \dot{S}_2 + \dot{S}_3 + \dot{S}_4$$
$$= -\frac{1}{m}|\delta_1|^2 - |\delta_2|^2 - |\delta_3|^2 - |\epsilon_3|^2 - |\delta_4|^2 - |\epsilon_4|^2$$

Recall that δ_1 and ϵ_3 together form the original tracking error that we wish to minimize. Thus, stabilizing the Lyapunov function must act to achieve the desired control objective.

15.5   Analysis

In this section, we give two results, which provide useful information on the evolution of the system given by (15.12)-(15.15) along with the control action given by (15.40), (15.41), (15.21) and (15.35).

The control design proposed in Section 15.4 provides a control law that guarantees robust trajectory tracking for the approximate model equations (15.12)-(15.15). The purpose of the analysis is to show that as long as the small body forces are sufficiently small, this control will ensure the same properties for the full system to within a neighbourhood of zero. Of key importance in the development is an understanding of the underlying error dynamics. Due to the choices made in the control law design, the error dynamics are linear dynamics perturbed by the small body forces. Let

$$\alpha = (\delta_1, \delta_2, \delta_3, \delta_4, \epsilon_3, \epsilon_4) \in \mathbb{R}^{14}$$

It can be verified that the error dynamics are given by

$$\dot{\alpha} = \begin{pmatrix} -\frac{1}{\eta}I_3 & \frac{1}{m}I_3 & 0 & 0 & 0 & 0 \\ -\frac{1}{m}I_3 & -I_3 & I_3 & 0 & 0 & 0 \\ 0 & -I_3 & -I_3 & I_3 & 0 & 0 \\ 0 & 0 & -I_3 & -I_3 & 0 & 0 \\ 0 & 0 & 0 & 0 & -1 & 1 \\ 0 & 0 & 0 & 0 & -1 & -1 \end{pmatrix} \alpha$$

$$+ \begin{pmatrix} 0 \\ R(\eta)Kw \\ \frac{m+1}{m}R(\eta)Kw \\ \frac{2m^2+2m+1}{m^2}R(\eta)Kw \\ 0 \\ 0 \end{pmatrix} \qquad (15.42)$$

Note that there are no perturbation terms in the last two block entries of the α dynamics and that these correspond to the yaw control for the orientation of ϕ. In fact, these terms are not particularly important in the following analysis, since the control design for these terms is mostly decoupled from the remaining errors.

In the following analysis, bounds made up of norms of the errors $|\delta_i|$, $i = 1, \ldots, 4$ and $|\epsilon_3|$ and $|\epsilon_4|$ are used regularly. To simplify notation, define

$$\gamma = (|\delta_1|, |\delta_2|, |\delta_3|, |\delta_4|, |\epsilon_3|, |\epsilon_4|) \in \mathbb{R}^6$$

and

$$\chi = (|\dot{\hat{v}}|, |\hat{v}^{(2)}|, |\hat{v}^{(3)}|, |\dot{\hat{\phi}}|, |\hat{\phi}^{(2)}|) \in \mathbb{R}^5$$

Thus a bound linear in the absolute norms of the error terms may be written as $\pi^T \gamma + \tau^T \chi$ for $\pi \in \mathbb{R}^6$, $\tau \in \mathbb{R}^5$, constant vectors. The Lyapunov function $\mathcal{L}$ may be written as

$$\mathcal{L} = \frac{1}{2}|\gamma|^2 \qquad (15.43)$$

Due to the presence of the small body forces in the error equations (15.42), then for the evolution of the true closed-loop system, the Lyapunov function may not be monotonically decreasing. Of course if $|w|$ is small relative to the errors γ, then one expects that the linear dynamics in (15.42) will dominate the perturbations and the Lyapunov function will be decreasing. The following lemma provides an analogous result to that obtained for the linearization method of control design [35, 48].

Lemma 15.1 *Consider the dynamics defined by (15.12)-(15.15). Let the controls $\tilde{w}$ and $\tilde{u}$ be given by (15.40) and (15.41) and recover the control inputs w and u from (15.21) and (15.35). Then, the Lyapunov function (15.43) is strictly decreasing as long as*

$$|\gamma|^2 \geq \sigma |w| \langle \pi_0, \gamma \rangle \qquad (15.44)$$

where

$$\pi_0 = \left(0, 1, \frac{m+1}{m}, \frac{2m^2 + m + 1}{m^2}, 0, 0 \right) \quad and \quad \sigma = \|K\|_2$$

and L and K are constant matrices given by the physical parameters of the helicopter (cf. [62]).   ∎

Proof 15.1 *Taking the derivative of $\mathcal{L}$, substituting from (15.42) and using Hölders inequality to bound the effect of the small body forces, yields*

$$\dot{\mathcal{L}} \leq -|\gamma|^2 + |\delta_2| \|R(\eta)Kw\| + \frac{m+1}{m} |\delta_3| \|R(\eta)Kw\|$$
$$+ \frac{2m^2 + m + 1}{m} |\delta_4| \|R(\eta)Kw\|$$
$$= -|\gamma|^2 - |\epsilon_3|^2 - |\epsilon_4|^2 + \sigma |w| \pi_0^T \gamma$$

where π_0 is given in the lemma statement. The result follows directly from this inequality.   ∎

Remark 15.2 *The constant $\sigma = \|K\|_2$ measures the inverse of the effective offset between the centre of mass of the helicopter and the center of the rotor disk (at which point the force u is applied). Thus, σ large corresponds to a small offset and correspondingly large "small body forces". It is clear that in such situations the approach taken, where the small body forces are considered as perturbations, becomes increasingly difficult to solve. For a model helicopter, a value of $\sigma \approx 2 - 5$ is obtained, corresponding to an offset of around 50cm-20cm. In contrast, for a full scale helicopter, the offset may be around 2m given a value of $\sigma \approx 0.5$.*   ∎

15.6   Simulations

In this section, simulations for two experiments are presented. The first experiment considered the case of stabilization of the helicopter dynamics to a stationary configuration given an initial offset. The second considered the case of trajectory tracking. The trajectory chosen for the tracking simulations was a helix ascending in the vertical direction.

The parameters for the helicopter used are given in Table 15.1. These values are based on measurements of the model helicopter used by the Swiss research group at the Measurement and Control Laboratory, ETH, Switzerland, along with estimates drawn from the literature.

Parameter	Value
Mass	18.085 kg
I_1^a	1.667 kg m^2
I_2^a	2.341 kg m^2
I_3^a	1.197 kg m^2
$|Q_M|$	0.02
$|Q_T|$	c0.002
l_T	(-1.5, 0, 0) m
l_M	(0, 0, -0.45) m
κ	0.42
g	9.80 m s^{-2}
σ	2.437

Table 15.1: Parameters of helicopter used

For all simulations, the following choice of the initial conditions was adopted, considering the initial translational dynamics control $u_0 = gm \approx 177$. Thus, the initial force input should be exactly that which sustains the helicopter in stationary flight. The initial condition chosen was

$$\xi_0 = \dot{\xi}_0 = \begin{pmatrix} 0 \\ 0 \\ 0 \end{pmatrix}, \quad \phi_0 = \dot{\phi}_0 = 0$$

In the first experiment, the regulation of position or stabilization of the helicopter was considered. The desired position $\hat{\xi}_0$ and $\hat{\eta}_0$ were chosen to be

$$\hat{\xi}_0 = \begin{pmatrix} 1 \\ 2 \\ -1 \end{pmatrix}, \quad \hat{\phi}_0 = 1 \text{ rad}$$

Results are shown in Figures 15.4 and 15.5 and were obtained by considering the complete model including small body forces. Figure 15.4

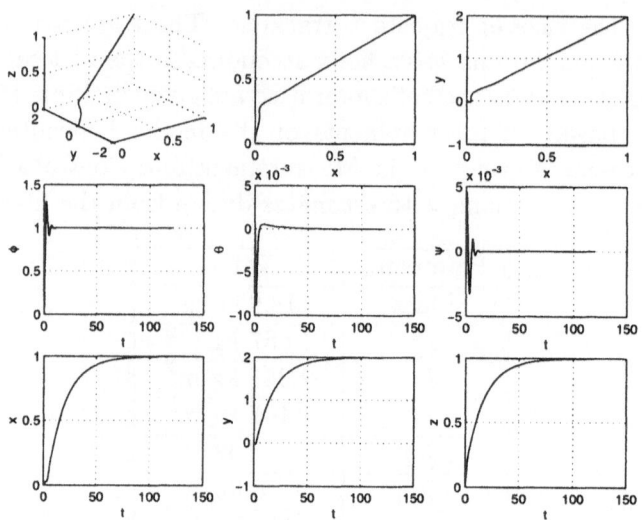

Figure 15.4: Position regulation of the helicopter dynamics in the presence of small body forces

illustrates the behavior of the helicopter. Figure 15.5 shows the decrease of the Lyapunov function.

The second experiment concerned the tracking problem. In this case, the desired trajectory was chosen as a helix ascending from a point

$$\hat{\xi}_0 = \begin{pmatrix} 1 \\ 2 \\ -1 \end{pmatrix}, \quad \hat{\phi}_0 = 1rd$$

that lies above, and to the side, of the initial position of the helicopter. The velocity of the desired trajectory was

$$\dot{\hat{\xi}} = \begin{pmatrix} 0.1\cos\hat{\phi} \\ 0.1\sin\hat{\phi} \\ -0.15 \end{pmatrix}, \quad \dot{\hat{\phi}}_0 = 0.1$$

Results are presented in Figures 15.6 and 15.7 for the full model of the helicopter including small body forces. In the same way, Figure 15.7 is presented to illustrate the decrease of the Lyapunov function.

The simulations indicated that the position regulation (to the desired trajectory) is achieved after a short transient of 50 and 100s (cf. Figure 15.4). The initial response of the perturbed system to the control action (when the small body forces are at their largest) shows noticeable

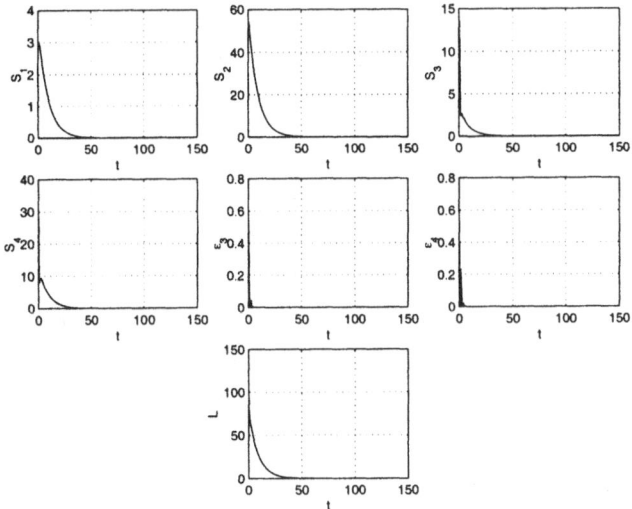

Figure 15.5: The Lyapunov function and its components for position regulation in the presence of small body forces

effects of the disturbances. In contrast, during the later stages of the convergence, the offset of the orientation angles θ and ψ is negligible. These angles, though not exactly representing the small body forces, are closely linked to the torque input to the system. Thus, for small angles θ and ψ it follows that the small body forces are negligible.

In the case of tracking problem (cf. Figure 15.6), the asymptotic effect of the small body forces appears to be negligible. This can be seen in the apparent convergence of the angles θ and ψ in Figure 15.6. In fact, these angles continue to oscillate but the scale of Figure 15.6 does not show these oscillations in comparison to the transients associated with the initial offset. Thus, the simulation illustrated by Figure 15.6 shows once more the relative insignificance of the small body force effects when one considers relatively slow desired trajectories.

Remark 15.3 *In the case where the damping effects are much less apparent, for example if the helicopter mass was an order of 10 times lighter, then it is possible to see the oscillatory nature of the zero dynamics appearing in the simulations. This is shown in a companion paper [62].* ∎

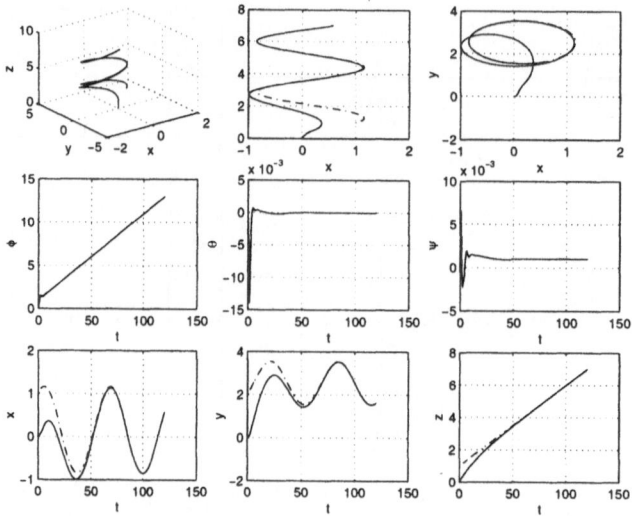

Figure 15.6: Behavior of the complete helicopter dynamics for the tracking problem (including small body forces)

15.7 Conclusions

In this chapter, a simple model for the dynamics of a helicopter close to hover conditions has been derived. It was shown that this model may be written in a form that is amenable to modern non-linear control design techniques. A backstepping control was designed based on an approximation of the system obtained by ignoring the small body forces associated with the torque control. An analysis was presented to show that the Lyapunov function is monotonically decreasing for large errors in the control objective.

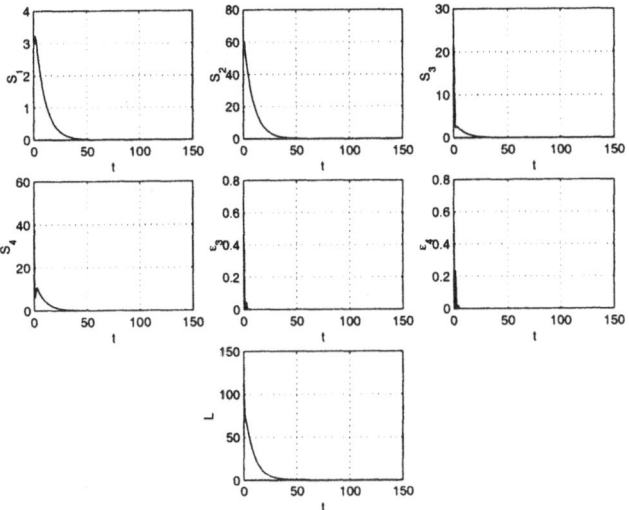

Figure 15.7: The Lyapunov function with its components, for the tracking problem considering the complete helicopter dynamics

Bibliography

[1] H. Arai and S. Tachi. Position control of a manipulator with passive joints using dynamic coupling. *IEEE Trans. Robotics and Automation*, 7(4):528–534, 1991.

[2] H. Arai, K. Tanie, and N. Shiroma. Nonholonomic control of a three-dof planar underactuated manipulator. *IEEE Trans. Robotics and Automation*, 14(5):681–695, 1998.

[3] S. Arimoto and T. Nakayama. Another language for describing motions of mechatronics systems: A nonlinear position-dependent circuit theory. *IEEE/ASME Transactions on Mechatronics*, 1(2), June 1996.

[4] K.J. Aström and K. Furuta. Swinging up a pendulum by energy control. *Automatica*, 36(2):287–295, February 2000.

[5] Ahmad N. Atassi and Hassan K. Khalil. A separation principle for the stabilization of a class of nonlinear systems. *IEEE Trans. on Automatic Control*, 44(9):1672–1687, 1999.

[6] J.C. Avila-Vilchis and B. Brogliato. Nonlinear passivity-based control for a scale model helicopter. In *26th European Rotorcraft Forum*, volume CD-ROM paper 21, The Hague, The Netherlands, 2000.

[7] J.C. Avila-Vilchis, B. Brogliato, and R. Lozano. Modélisation d'hélicoptère. Technical Report AP 00-021, Laboratoire d'Automatique de Grenoble, INPG, France, 2000.

[8] J. Baillieul. Kinematically redundant robots with flexible components. *IEEE Control Systems Magazine*, 13:15–21, 1993.

[9] S.P. Berge, K. Ohtsu, and T.I. Fossen. Nonlinear tracking control of underactuated ships minimizing the cross-track error. In

Proceedings of the IFAC Conference on Control Applications in Marine Systems (CAMS'98), pages 141–147, Fukuoka, Japan, October 27-30, 1998.

[10] D. J. Block. Mechanical design and control of the pendubot. Master's thesis, University of Illinois, Urbana-Champaign, IL, USA, 1996.

[11] S. A. Bortoff. *Pseudolinearization using Spline Functions with Application to the Acrobot.* PhD thesis, Dept. of Electrical and Computer Engineering, University of Illinois, Urbana-Champaign, IL, USA, 1992.

[12] R. Bradley. The flying brick exposed: nonlinear control of a basic helicopter model. Technical Report TR/MAT/RB/6, Department of Mathematics, Glasgow Caledonian University, Scotland, U.K., 1996.

[13] R. W. Brockett. Asymptotic stability and feedback stabilization. In R.S. Millman R.W. Brockett and H.J. Sussmann, editors, *Differential Geometric Control Theory*, pages 181–191. Birkhäuser, 1983.

[14] B. Brogliato, R. Lozano, and I. D. Landau. New relationships between Lyapunov functions and the passivity theorem. *International Journal Adaptive Control and Signal Processing*, 7:353–365, 1993.

[15] F. Bullo and N. E. Leonard. Motion primitives for stabilization and control of underactuated vehicles. In *Preprints of the 4th IFAC NOLCOS'98*, volume 1, pages 133–138, Enschede, The Netherlands, 1998.

[16] C.I. Byrnes, A. Isidori, and J.C. Willems. Passivity, feedback equivalence and the global stabilization of nonminimum phase nonlinear systems. *IEEE Transactions on Automatic Control*, 36:1228–1240, 1991.

[17] C.C. Chung and J. Hauser. Nonlinear control of a swinging pendulum. *Automatica*, 31(6):851–862, 1995.

[18] J. Collado, R. Lozano, and I. Fantoni. Control of convey-crane based on passivity. In *Proceedings of the American Control Conference ACC'00*, Chicago, USA, June 2000.

[19] I. Fantoni and R. Lozano. Stabilization of the Furuta pendulum around its homoclinic orbit. In *Preprints of the 5ᵗʰ IFAC NOLCOS'01*, St. Petersburg, Russia, 2001.

[20] I. Fantoni, R. Lozano, and F. Mazenc. Control of the PVTOL aircraft using the forwarding technique and a Lyapunov approach. In *Proceedings of the European Control Conference ECC'01*, Porto, Portugal, 2001.

[21] I. Fantoni, R. Lozano, F. Mazenc, and A. M. Annaswamy. Stabilization of a two-link robot using an energy approach. In *Proceedings of the European Control Conference ECC'1999*, Karlsruhe, Germany, 1999.

[22] I. Fantoni, R. Lozano, F. Mazenc, and K. Y. Pettersen. Stabilization of a nonlinear underactuated hovercraft. *International Journal of Robust and Nonlinear Control*, 10:645–654, 2000.

[23] I. Fantoni, R. Lozano, and M. W. Spong. Stabilization of the reaction wheel pendulum using an energy approach. In *Proceedings of the European Control Conference ECC'01*, Porto, Portugal, 2001.

[24] I. Fantoni, R. Lozano, and M. W. Spong. Energy based control of the pendubot. *IEEE Transactions on Automatic Control*, 45(4):725–729, April 2000.

[25] T. I. Fossen. *Guidance and Control of Ocean Vehicles*. Chichester: John Wiley & Sons Ltd, 1994.

[26] T. I. Fossen, J.-M. Godhavn, S. P. Berge, and K.-P. Lindegaard. Nonlinear control of underacuated ships with forward speed compensation. In *Preprints of the 4ᵗʰ IFAC NOLCOS'98*, volume 1, pages 121–126, Enschede, The Netherlands, 1998.

[27] A. L. Fradkov. Swinging control of nonlinear oscillations. *Int. J. Control*, 64(6):1189–1202, 1996.

[28] E. Frazzoli, M. Dahleh, and E. Feron. Trajectory tracking control design for autonomous helicopters using a backstepping algorithm. In *American Control Conference ACC'00*, Chicago, USA, June 2000.

[29] R. A. Freeman and P. V. Kokotović. *Robust Nonlinear Control Design: State-space and Lyapunov techniques*. Systems and Control: Foundations and Applications. Birkhäuser, Boston, 1996.

[30] K. Furuta, M. Yamakita, and S. Kobayashi. Swing-up control of inverted pendulum using pseudo-state feedback. *Journal of Systems and Control Engineering*, 206(6):263–269, 1992.

[31] R. Ghanadan. Nonlinear control system design via dynamic order reduction. In *Proc. of the Conference on Decision and Control*, pages 3752–3757, Florida, USA, 1994.

[32] J.-M. Godhavn. *Topics in Nonlinear Motion Control*. PhD thesis, Department of Engineering Cybernetics, Norwegian University of Science and Technology, Trondheim, Norway, 1997.

[33] H. Goldstein. *Classical Mechanics*. Addison-Wesley Series in Physics. Addison-Wesley, U.S.A., second edition, 1980.

[34] J.-T. Gravdahl and O. Egeland. *Compressor Surge and Rotating Stall: Modeling and Control*. Springer-Verlag London, 1990.

[35] J. Hauser, S. Sastry, and G. Meyer. Nonlinear control design for slightly nonminimum phase systems: Application to V/STOL aircraft. *Automatica*, 28(4):665–679, 1992.

[36] D.J. Hill. Preliminaries on passivity and gain analysis. *IEEE CDC Tutorial Workshop on Nonlinear Controller Design using Passivity and Small-Gain Techniques*, 1994.

[37] D.J. Hill and P.J. Moylan. Stability of nonlinear dissipative systems. *IEEE Transactions on Automatic Control*, AC-21:708–711, 1976.

[38] D.J. Hill and P.J. Moylan. Stability results for nonlinear feedback systems. *Automatica*, 13:377–382, 1977.

[39] D.J. Hill and P.J. Moylan. Connections between finite gain and asymptotic stability. *IEEE Transactions on Automatic Control*, AC-25:931–936, 1980.

[40] D.J. Hill and P.J. Moylan. Dissipative dynamical systems:basic input-output and state properties. *J. Franklin Inst.*, 309:327–357, 1980.

[41] R.A. Horn and C.R. Johnson. *Matrix Analysis*. Cambridge University Press, 1985.

[42] A. Isidori. *Nonlinear Control Systems: An Introduction.* 3rd ed. Springer-Verlag Berlin, 1995.

[43] M. Iwashiro, K. Furuta, and K. J. Åström. Energy based control of pendulum. In *Proceedings of the 1996 IEEE International Conf. on Control Applications*, pages 715–720, 1996.

[44] E. Atlee Jackson. *Perspectives of Nonlinear Dynamics.* Cambridge University Press, 1989.

[45] B. Jakubczyk and W. Respondek. On the linearization of control systems. *Bull. Acad. Polon. Sci. Math.*, 28:517–522, 1980.

[46] H. K. Khalil. *Non-Linear Systems.* Prentice Hall, Second Edition, 1996.

[47] T. J. Koo, F. Hoffmann, H. Shim, and S. Sastry. Control design and implementation of autonomous helicopter. In *Proceedings of the 37th IEEE Conference on Decision and Control (CDC'98)*, Florida, USA, 1998.

[48] T. J. Koo and S. Sastry. Output tracking control design of a helicopter model based on approximate linearization. In *Proceedings of the 37th IEEE Conference on Decision and Control (CDC'98)*, Florida, USA, 1998.

[49] M. Krstić, I. Kanallakopoulos, and P. V. Kokotović. *Nonlinear and Adaptative Control Design.* Wiley, New York, 1995.

[50] E. Lefeber. *Tracking Control of Nonlinear Mechanical Systems.* PhD thesis, Faculty of Mathematical Sciences, University of Twente, Enschede, The Netherlands, 2000.

[51] N. E. Leonard. Periodic forcing, dynamics and control of underactuated spacecraft and underwater vehicles. In *Proc. 34th IEEE Conf. on Decision and Control*, pages 3980–3985, New Orleans, LA, 1995.

[52] W. S. Levine. *The Control Handbook.* CRC Press in cooperation with IEEE Press, 1996.

[53] E. Liceaga-Castro, R. Bradley, and R. Castro-Linares. Helicopter control design using feedback linearization techniques. In *Proceedings of the 28th Conference on Decision and Control, CDC'89*, pages 533–534, Tampa, FL, USA, 1989.

[54] F. Lin, W. Zhang, and R. D. Brandt. Robust hovering control of a PVTOL aircraft. *IEEE Transactions on Control Systems Technology*, 7(3):343–351, 1999.

[55] Z. Lin, A. Saberi, M. Gutmann, and Y. A. Shamash. Linear controller for an inverted pendulum having restricted travel: A high-and-low approach. *Automatica*, 32(6):933–937, 1996.

[56] R. Lozano, B. Brogliato, O. Egeland, and B. Maschke. *Dissipative Systems Analysis and Control: Theory and Applications.* Springer-Verlag, Communications and Control Engineering Series, London, 2000.

[57] R. Lozano, B. Brogliato, and I. D. Landau. Passivity and global stabilization of cascaded nonlinear systems. *IEEE Transactions on Automatic Control*, 37(9):1386–1388, 1992.

[58] R. Lozano and I. Fantoni. Passivity based control of the inverted pendulum. In *Preprints of the 4th IFAC NOLCOS'98*, volume 1, pages 145–150, Enschede, The Netherlands, 1998.

[59] R. Lozano, I. Fantoni, and D. J. Block. Stabilization of the inverted pendulum around its homoclinic orbit. *Systems & Control Letters*, 40(3):197–204, 2000.

[60] De Luca and Siciliano. Regulation of flexible arms under gravity. *IEEE Trans. on Robotics and Automation*, 9(4):463–467, 1993.

[61] R. Mahony and T. Hamel. Robust trajectory tracking for a scale model autonomous helicopter. *Submitted to International Journal on Robotics Research*, 1999.

[62] R. Mahony, T. Hamel, and A. Dzul. Hover control via Lyapunov control for an autonomous model helicopter. In *Proceedings of the 38th IEEE Conf. on Decision and Control*, Phoenix, Arizona, USA, 1999.

[63] R. Mahony and R. Lozano. An energy based approach to the regulation of a model helicopter near to hover. In *Proceedings of the European Control Conference ECC'1999*, Karlsruhe, Germany, 1999.

[64] R. Mahony and R. Lozano. (Almost) exact path tracking control for an autonomous helicopter in hover manoeuvres. In *International Conference on Robotics and Automation, ICRA2000*, San Fransisco, California, USA, 2000.

[65] A. Makhlin. Design and control of an inverted pendulum system. Master's thesis, University of Illinois, Urbana-Champaign, IL, USA, 1998.

[66] P. Mallhaupt, B. Srinivasan, J. Levine, and D.Bouvin. A toy more difficult to control than the real thing. In *Proceedings of the European Control Conference, ECC'97*, Brussels, Belgium, July 1997.

[67] P. Martin, S. Devasia, and Brad Paden. A different look at output tracking: Control of a VTOL aircraft. *Automatica*, 32(1):101–107, 1996.

[68] B.M. Maschke, A.J. van der Schaft, and P.C. Breedveld. An intrinsic hamiltonian formulation of network dynamics: Non-standard poisson structures and gyrators. *J. Franklin Inst.*, 329:923–966, 1992.

[69] F. Mazenc. *Stabilisation de trajectoires, ajout d'intégration, commande saturées*. PhD thesis, Ecole des Mines de Paris, 1996.

[70] F. Mazenc, A. Astolfi, and R. Lozano. Lyapunov function for the ball and beam: Robustness property. In *Proceedings of the 38th IEEE Conf. on Decision and Control*, Phoenix, Arizona, USA, Dec. 1999.

[71] F. Mazenc, R. Mahony, and R. Lozano. Forwarding control of reduced scale autonomous helicopter: A Lyapunov control design. *Draft version*.

[72] F. Mazenc and L. Praly. Adding integrations, saturated controls, and stabilization for feedforward systems. *IEEE Transactions on Automatic Control*, 41(11):1559–1578, November 1996.

[73] R. M. Murray, M. Rathinam, and M. van Nieuwstadt. An introduction to differential flatness of mechanical systems. In *Proceedings of "Ecole d'Eté, Théorie et pratique des systèmes plats"*, Grenoble, France, 1996.

[74] Y. Nakamura, T. Suzuki, and M. Koinuma. Nonlinear behavior and control of a nonholonomic free-joint manipulator. *IEEE Trans. on Robotics and Automation*, 13(6):853–862, 1997.

[75] H. Nijmeijer and A. J. van der Schaft. *Nonlinear Dynamical Control Systems*. Springer-Verlag, 1990.

[76] R. Olfati-Saber. *Nonlinear Control of Underactuated Mechanical Systems with Application to Robotics and Aerospace Vehicles*. PhD thesis, Department of Electrical Engineering and Computer Science of the Massachusetts Institute of Technology, Cambridge, USA, 2001.

[77] R. Olfati-Saber. Fixed point controllers and stabilization of the cart-pole system and the rotating pendulum. In *Proceedings of the 38th IEEE Conf. on Decision and Control*, pages 1174–1181, Phoenix, Arizona, USA, Dec. 1999.

[78] R. Olfati-Saber. Cascade normal forms for underactuated mechanical systems. In *Proceedings of the 39th IEEE Conf. on Decision and Control*, Sydney, Australia, Dec. 2000.

[79] R. Olfati-Saber. Global configuration stabilization for the VTOL aircraft with strong input coupling. In *Proceedings of the 39th IEEE Conf. on Decision and Control*, Sydney, Australia, Dec. 2000.

[80] G. Oriolo and Y. Nakamura. Control of mechanical systems with second-order nonholomic constraints: Underactuated manipulators. In *Proc. 30th IEEE Conf. on Decision and Control*, pages 2398–2403, Brighton, England, 1991.

[81] K. Y. Pettersen. *Exponential stabilization of underactuated vehicles*. PhD thesis, Department of Engineering Cybernetics, Norwegian University of Science and Tecnology, Trondheim, Norway, 1996.

[82] K. Y. Pettersen and O. Egeland. Exponential stabilization of an underactuated surface vessel. In *Proceedings 35th IEEE Conf. on Decision and Control*, pages 967–971, 1996.

[83] K. Y. Pettersen and T. I. Fossen. Underactuated ship stabilization using integral control: experimental results with cybership i. In

Preprints of the 4th IFAC NOLCOS'98, volume 1, pages 127–132, Enschede, The Netherlands, 1998.

[84] K. Y. Pettersen and H. Nijmeijer. Tracking control of an under-actuated surface vessel. In *Proc. 37th IEEE Conf. on Decision and Control, CDC 98*, pages 4561–4566, December 1998.

[85] K. Y. Pettersen and H. Nijmeijer. Global practical stabilization and tracking for an underactuated ship - a combined averaging and backstepping approach. In *Proc. IFAC Conference on System Structure and Control*, pages 59–64, July 1998.

[86] L. Praly. Stabilisation du système pendule-chariot: Approche par assignation d'énergie. *personal communication*, 1995.

[87] R. W. Prouty. *Helicopter Performance, Stability and Control.* Krieger Publishing Company, 1995.

[88] M. Reyhanoglu, A. J. van der Schaft, N. H. McClamroch, and I. Kolmanovsky. Dynamics and control of a class of underactuated mechanical systems. *IEEE Transactions on Automatic Control*, 44(9):1663–1671, 1999.

[89] Ch. Rui, I. Kolmanovsky, and P.J. McNally. Attitude control of underactuated multibody spacecraft. In *IFAC, 13th Triennial World Congress*, pages 425–430, San Francisco, USA, 1996.

[90] W. Schaufelberger and H. Geering. Case study on helicopter control. In *Control of Complex Systems (COSY) Symposium (Invited session)*, Macedonia, October, 1998.

[91] R. Sepulchre, M. Janković, and P. Kokotović. *Constructive Nonlinear Control.* Springer-Verlag London, 1997.

[92] D. Seto and J. Baillieul. Control problems in super-articulated mechanical systems. *IEEE Transactions on Automatic Control*, 39:2442–2453, 1994.

[93] O. Shakernia, Y. Ma, T. Koo, and S. Sastry. Landing an unmanned air vehicle: Vision based motion estimation and nonlinear control. *Asian Journal of Control*, 1(3):128–145, 1999.

[94] I. H. Shames. *Engineering Mechanics: Dynamics.* Prentice Hall International, 1967.

[95] H. Shim, T. Koo, F. Hoffmann, and S. Sastry. A comprehensive study of control design for an autonomous helicopter. In *Proceedings of the 37th Conference on Decision and Control CDC'98*, 1998.

[96] A. S. Shiriaev. Control of oscillations in affine nonlinear systems. In *Proc. IFAC conference "System Structure and Control"*, pages 789–794, Nantes, France, 1998.

[97] A. S. Shiriaev. The notion of v-detectability and stabilization of invariant sets of nonlinear systems. In *Proc. 37th IEEE Conf. on Decision and Control, CDC 98*, pages 2509–2514, Tampa, FL, USA, December 1998.

[98] A. S. Shiriaev, O. Egeland, and H. Ludvigsen. Global stabilization of unstable equilibrium point of pendulum. In *Proceedings of the 37th IEEE Conf. on Decision and Control, CDC 98,*, Tampa, FL, USA, 1998.

[99] A. S. Shiriaev and A. L. Fradkov. Stabilization of invariant manifolds for nonlinear non-affine systems. In *Preprints of the 4th IFAC NOLCOS'98*, pages 215–220, Enschede, The Netherlands, 1998.

[100] A. S. Shiriaev, A. Pogromsky, H. Ludvigsen, and O. Egeland. On global properties of passivity-based control of an inverted pendulum. *International Journal of Robust and Nonlinear Control*, 10:283–300, April 2000.

[101] A. Sira-Ramirez, R. Castro-Linares, and E. Licéaga-Castro. Regulation of the longetudinal dynamics of an helicopter system: A Liovillian systems approach. In *Proceedings of the American Control Conference ACC'99*, San Diego, California, USA, 1999.

[102] A. Sira-Ramirez, R. Castro-Linares, and E. Licéaga-Castro. A Liouvillian systems approach for the trajectory planning-based control of helicopter models. *International Journal of Robust and Nonlinear Control*, 10:301–320, 2000.

[103] A. Sira-Ramirez, M. Zribi, and S. Ahmed. Dynamic sliding mode control approach for vertical flight regulation in helicopters. *IEE Proceedings: Control Theory and Applications*, 141(1):19–24, January 1994.

[104] H. Sira-Ramirez and R. Castro-Linares. On the regulation of a helicopter system: A trajectory planning approach for the Liouvillian model. In *Proceedings of the European Control Conference ECC'1999*, 1999.

[105] J. Slotine and W. Li. *Applied nonlinear control.* Prentice Hall, 1991.

[106] M. W. Spong. The swing up control of the acrobot. In *IEEE Int. Conf. on Robotics and Automation*, San Diego, CA, 1994.

[107] M. W. Spong and D. J. Block. The pendubot: A mechatronic system for control research and education. In *Proceedings of the 34th IEEE Conf. on Decision and Control*, 1995.

[108] M. W. Spong, P. Corke, and R. Lozano. Nonlinear control of the reaction wheel pendulum. To appear in *Automatica*, 2001.

[109] M.W. Spong and L. Praly. Control of underactuated mechanical systems using switching and saturation. In *Proceedings of the Block Island Workshop on Control Using Logic Based Switching*, 1996.

[110] M.W. Spong and M. Vidyasagar. *Robot Dynamics and Control.* John Wiley & Sons, Inc., 1989.

[111] W. Z. Stepniewsky. *Rotor-wing Aerodynamics, Vol. 1 Basic Theories of Rotor Aerodynamics.* Dover Publishing., Inc., N. Y., 1984.

[112] J. P. Strand, K. Ezal, T. I. Fossen, and P. V. Kokotovic. Nonlinear control of ships: a locally optimal design. In *Preprints of the 4th IFAC NOLCOS'98*, Enschede, The Netherlands, 1998.

[113] M. Takegaki and S. Arimoto. A new feedback method for dynamic control of manipulators. *Trans. ASME, J. Dyn. Systems, Meas. Control*, 103:119–125, 1981.

[114] F. Tchen-Fo, C. Allain, and A. Desopper. Improved vortex ring model for helicopter pitch up prediction. In *26th European Rotorcraft Forum*, volume CD-ROM paper 42, The Hague, The Netherlands, 2000.

[115] Andrew R. Teel. A nonlinear small gain theorem for the analysis of control systems with saturation. *IEEE Transactions on Automatic Control*, 41(9):1256–1270, 1996.

[116] D. Thomson and R. Bradley. Recent develoments in the calculation of inverse dynamic solutions of the helicotper equations of motion. In *Proceedings of the U.K. Simulation Council Triennial Conference*, 1987.

[117] A. J. van der Schaft. L_2-*Gain and Passivity Techniques in Nonlinear Control*. Springer-Verlag, 1996.

[118] M. van Nieuwstadt and J. Morris. Control of rotor speed for a model helicopter: A design cycle. In *Proceedings of the American Control Conference*, pages 688–692, 1995.

[119] M. van Nieuwstadt and R. Murray. Outer flatness: Trajectory generation for a model helicopter. In *Proceedings of the European Control Conference ECC'97*, Brussels, Belgium, 1997.

[120] F. Verduzco and J. Alvarez. *Stability and Bifurcations of an Underactuated Robot Manipulator*. CICESE, Mexico.

[121] Q. Wei, W.P. Dayawansa, and W.S. Levine. Nonlinear controller for an inverted pendulum having restricted travel. *Automatica*, 31(6):841–850, 1995.

[122] J.C. Willems. Dissipative dynamical systems - part 1: General theory. *Arch. Rational Mechanics and Analysis*, 45:321–351, 1972.

[123] M. Yamakita, M. Iwashiro, Y. Sugahara, and K. Furuta. Robust swing up control of double pendulum. In *Proceedings of the American Control Conference*, pages 290–295, Seattle, Washington, 1995.

Index